GROWING UP
AT THE
DESERT QUEEN RANCH

Library of Congress Catalog Card No. 97-78014

ISBN 0-9617961-6-2

This printing by:

Artcraft Printers
Billings, Montana

Third Printing - 2008

DEDICATION

Nelda and Harmon King

Volunteer caretakers at the Desert Queen Ranch, who for seven years (1991 - 1998) protected this historic resource, helped with equipment and building stablization, watered and cared for the orchard trees, gave interpretive tours, and made Joshua Tree National Park visitors feel welcome. For their untiring service to the ranch and to the National Park Service, this book is dedicated to them in grateful appreciation.

TABLE OF CONTENTS

ACKNOWLEDGEMENTS

Several individuals have contributed to the creation of this book, and we would like to thank the following: Corinne Keys for proofreading; Walt Bolster, Ross Carmichael, Lucille Frasher, Nelda and Harmon King, and Mel Weinstein for providing photos which are individually acknowledged where used. All uncredited current-day ranch photos were taken by Art Kidwell, and most of the historic photos, unless credited, came from his collection. Vickie Waite of *The Sun Runner* Magazine in 29 Palms scanned the photos into the text pages. Her help, suggestions, and expertise were invaluable in creating the camera-ready pages. The help of all of these contributors is much appreciated.

Former Joshua Tree National Park Superintendent Dave Moore, now retired, enjoys hearing Willis Keys tell about growing up at the Desert Queen Ranch in October 1993. Photo by Art Kidwell.

INTRODUCTION

In a rock-enclosed canyon in Joshua Tree National Park only a few miles from Hidden Valley campground, a haven for rock climbers, lies the Desert Queen Ranch. This location has been a living site for early native American people, a cattle ranch, a gold ore milling site, and the homestead home and ranch of Bill Keys and his family.

Bill arrived here in 1910 when Frederick B. Morgan, a mining promoter from Pasadena owned the Desert Queen Mine and the millsite here in the canyon. Bill worked for him as a caretaker and as a assayer at the mine. When Morgan died, his widow suggested that Bill file on both for back wages owed him. In 1917 he homesteaded additional acreage adjoining the 5-acre millsite and the total 160 acres became the Desert Queen Ranch. In 1918 he brought Frances Mae Lawton from Los Angeles as his bride.

Willis Keys, his brother Ellsworth, and sisters Virginia, Pat, and Phyllis grew up at the ranch. Since the nearest town in those early days was Banning, almost 60 miles away, the family learned to be conscien-tious and resourceful in living in the then-isolated desert area. Knowing the value of water in the harsh environment, they created dams to impound rain water for drinking, cooking, as well as irrigation. The family worked together as a team to accomplish the many chores necessary to insure the family's survival. Growing up here, there was "no time for play."

This collaborative book is a result of the co-authors' close friendship and their desire for ranch visitors and future Park Service rangers leading its guided tours to have a better understanding of what daily life was like growing up at the Desert Queen Ranch.

Willis' keen ability to remember the smallest details of past events in his life have been a great help in documenting not only the history of the ranch but also the early days of Joshua Tree National Park. The stories selected for this book were taken from 15 taped conversations, interviews, and walks through the ranch between the co-authors over a 22-year period. Other chapters were written from a historian's perspective to provide more indepth information and background for some of the stories.

Art Kidwell and Willis Keys

Photo of Bill Keys in the late 1930s taken by Ruth Schneider.

DAD

As far back as I can remember, my dad didn't have much hair. In his earlier days, he wore his hair long and used to part it down the middle. Around 1900 it was shoulder-length and really dark brown. One time when it was very long, he had it cut and made into a wig. My sister used to have it.

Dad said that he started to lose his hair after a trip up into the High Sierras. He and a partner went up to the Coso Hot Springs area to prospect. While he was there he got typhoid fever and almost died. He said that after that his hair started to fall out. In his later years and by the time of the Bagley shooting, all he had was a fringe like mine or a little bit more. Then it was coal black.

My dad was a rather patient person. When we were young, he tried to impart as much of his store of knowledge to us as possible. Once he told us or showed us how to do something, he expected us to retain that knowledge. If we didn't, he would be put-out.

When I was old enough to do some chores around the ranch, I would do one thing and wait for my dad to tell me what to do next. That didn't last very long. He said, "Don't wait to be told everything you do. You can see what needs to be done or what should be done and go ahead and do it!

I have remembered that over the years, and I feel that it has served me well. I have worked for many organizations big and small and so many times have seen a worker finish a job and wait for the boss to tell him what to do next. Well, I always remembered what my dad said and found something to do. Sometimes it wasn't what the boss had in mind, but I was always busy.

My dad taught us many things. He had a vast knowledge of nature and enjoyed explaining everything from the stars and the universe to animals and mineralogy. He also had a great knowledge of how to do things. Having to work alone most of the time, he devised ways to move heavy machinery and other things with rollers, pry bars, block and tackle, and chain block and tripod.

One time dad sold a stamp mill to a fellow and agreed to get the mill out of the Desert Queen Mine. The roads had washed out years before and had left large boulders in the canyon where the road was. So Dad built a rail line down the side of the mountain to the mill. He used a mine rail car and an old hoist to pull the loaded car up the hill. So he was able to move the mill, the crusher, the engine and the accompanying equipment in that way.

Dad was also knowledgeable in the art of blacksmithing and did a lot of work with it. He saved scrap iron for use in making things. Dad knew how to do most everything that had to be done around the ranch. He could raise a great garden and orchard, break horses and mules, do mining, use dynamite, run a stamp mill, do mechanical work, herd cattle, work with stone and cement, develop springs and build dams.

Dad could tell you the name of most all of the rocks and what minerals were in them just by examining them. He could tell what areas were good for prospecting just by looking at the formations.

He worked long hours to keep the ranch going. He watered the garden and orchard and took care of the animals - usually the two milk cows, two horses, some goats and chickens, and the cattle on the range.

There also was mining work to do. Dad had as many as thirty mining claims at one time. The government required that you do $100 worth of work on each claim each year as assessment work. So that required quite some time each year.

Dad milled ore from some of his claims and also for other people that were working the mines during the 1930's. He milled mostly for others, as there was quite a bit of mining activity out there during the Depression.

I can remember my dad taking me out to the orchard when I was about five or six years old to teach me how to shoot a .22. We had a lot of birds come and ruin the fruit, so Dad would take me out each day and give me a lesson in shooting birds. When he felt that I was capable of handling the .22, he would send me by myself after giving me six or eight bullets and telling me to bring back six or eight birds. So I'd go out there and spend an hour or two hunting for those birds.

When I was growing up, we didn't have a whole lot of ammunition so there wasn't a whole lot of target practice, but I got so I could shoot pretty good. And when I was old enough I could take off and go out and rabbit hunt, or my dad would take me once in awhile when he was hunting.

When I went out on my own to hunt rabbits, I had my dad's single shot Stevens. I remember when he gave it to me. It was just after the Homer Urton incident. Dad came home and told my mother what had happened. And he took me out to the barn and showed me what to feed to the animals. Then he handed the .22 to me and said, "That's yours. Now you'll have to take care of the place." So that's when I inherited the .22.

When I went into the service, I had a whole collection of .22 rifles. And they all went away. I think one of my sister's boy friends took them. I had about six or eight .22 that I had saved as a collection.

One time a fellow came up and stayed at that cabin by the lake for a week or so. He was out here on a vacation. And he had a little single shot .22 which opened up like a shotgun. It was a Hamilton.

So my dad and I were up there at the cabin talking with him, and he said, "Come on out and see what you can do with this rifle." So we did a little target practicing and I did pretty well. And he said, "I'll give you this rifle." And he gave it to me.

I remember in later years that rifle used to be advertised in the pulp magazines I think for $2.75. It was cheap, but it was a pretty nice rifle. And I had that and like a dummy, I made a pistol out of it. I cut the stock and made a pistol grip stock, and cut the barrel. Then I had a pistol. I don't know whatever happened to it.

Dad was an excellent shot with a rifle or pistol having practiced a lot in earlier years. Even when he was older, he would amaze me with his shooting ability.

He was very confident in everything that he did and gave confidence to others around him. Sometimes the going would be pretty tough, but it never stopped him. Dad would always find a way to get a task done.

Bill Keys examining bighorn sheep skull in February 1922.
Photo by Austin Armer.

MOTHER

My mother's maiden name was Frances Lawton, and she was born in Toledo, Ohio. Her father was a mining man, who was more in the promotion end of it, I believe. As a young child she moved with her family to Georgian Bay, Canada and then later to California. By then she had three brothers and three sisters. Mother worked in Los Angeles for a good many years as a stenographer and also for the Western Union. That's how she met my Dad.

Mother's three brothers - my uncles, Aaron, Albert, and Lansing (whom we called "Buster") were oil men who worked in the oil fields during the time that Long Beach was booming and later Bakersfield. They did come out here and work in the mines for a while during the '30s, and my one uncle, Aaron, homesteaded with his family up in Covington Flats at the head of Smithwater Canyon. But there was no way to make a living there, so he had to give it up.

We never talked back to our parents. That just was never allowed. At the table you were supposed to have good table manners. My mother instructed us very carefully, and my dad backed her up, of course. She was very understanding, but strict. She had been raised fairly strict so she passed it on to us.

She was somewhat religious, and her religion was Christian Science. Of course, out here with no opportunity to go to church or anything, she'd read the Bible and some of her literature. There was no Sunday School, but my mother did try to give us a little idea of what it was.

Bill and Frances with children: Willis, Virginia, and Ellsworth, about 1926.

She worked hard and always tried to keep things clean. She did all the housework, washed by hand, and canned during the canning season. She hardly ever had a spare moment, as there was always a lot of work to do.

Mother was an excellent cook. She used to do all of the cooking, and she did most of it from memory as far as recipes were concerned. She could remember just the right proportions, because it always came out good. And people coming through tried to make it here somewhere around mealtime, because they always knew they'd get a good meal. In the kitchen she had a wood stove which kept it warm in the winter and in the summer too. That made it a hot place to work.

When we didn't have beef because the stock wasn't ready to butcher, she'd fix cottontail or jackrabbit or chicken or something like that. My mother could really make some good meals out of those. She had a lot of her ideas that she used to prepare them.

We all got up early and did a good part of our chores before breakfast. Then we came in and had a good large bowl of oatmeal or cornmeal mush. She'd also fix pancakes, eggs, and bacon or possibly steaks with biscuits and gravy. So breakfast was a good large meal of the day, and something that when I got away from home, I found out wasn't the normal thing for most people.

Before too many of the family came along, Mother used to go with Dad to the mines and camp out while he did the assessment work or whatever was to be done. He worked several mines at different times, and then he'd haul the ore back and mill it. He'd bring it back by horse and wagon usually or mules and wagon. Mother was about 5' 3' and had blue eyes. When she was young, her hair was light brown. When she was older her hair was white.

My mother was very conscious of her skin and complexion. She never went out into the sun without a head cover which was usually a sun bonnet. She and my grandmother made many sun bonnets for themselves and to give as presents.

When my sister, Virginia and I were six and eight years old, my mother would take an hour about two days a week and hold a little school session. Though she had no teaching experience, she had a great store of knowledge. That instruction served us well when later we went to regular classes.

She was also a good instructor in practical knowledge. She taught Virginia and I how to sew. We made clothes for dolls and other things. She used to buy us one of those paper doll books where you can cut out the paper dolls and the dresses. When we ran out of clothes, we'd take an old Sears-Roebuck catalog which had the

Frances Keys holding a piece
of her purple glass collection
in the late 1950s.

colored pages of dresses and clothes, and cut them out and make tabs on them. That worked fine, and we had quite a collection.

I also remember her showing me how to gather the eggs and to look for hidden nests. The chickens had the run of their pen as well as the barn and corral area and would hide their nests in out of the way places. I was probably about three or four years old, but I can still remember going from nest to nest with a basket. There would be anywhere from six to twelve eggs in it when I finished every evening.

My mother had a large collection of colored glass, which she had saved over a number of years. In the old days a lot of the mines out there had bottles, in particular, tobacco jars. They used to pack tobacco in these round jars about six inches in diameter and about six inches high with a glass lid. They were nice little jars. Some of them were different shapes. Some of them were straight, but they took color. Well, she collected a lot of those. In fact she used to use them in the cupboards for sugar, salt, and other things quite a bit.

She collected other things that she came across in some of the old mining places. In earlier years at the ranch, I don't know if she actually put too many of them out in the sun. She just saved the color glass, then she got the idea of getting a table. My dad built her one, and she put all of that colored glass out there. And of course, it got better as the years went by. So she had quite a large collection of that stuff. She took an interest in a lot of things like that.

My grandmother would usually come out in the spring and stay all summer. She and my mother would work together quite a bit. My grandmother would help with the canning and around the house sewing. My grandmother liked to sew, so she and my mother would get together and sew patch-work quilts. They both would save material all year, then when she Grandmother came up to visit, she and my mother would cut this material up into little squares and make quilts and stuff them with some of the Angora wool from the goats. They were warm, and I think that each of us kids had one on our bed.

When we were young we got special instructions from the folks as to where we could go and not go around the ranch. We were to stay away from the well in the lower part of the orchard and the old mill area where there was another open well. We were also to stay away from the lake and the rocks other than those just around the house.

I being the oldest got special instructions to look after my sister and brother. Later when we were older and able to take care of ourselves, we could go rock climbing and do about anything that we wanted to do.

After graduating from high school, I went to work in Los Angeles and lived with Lee and Chet Perkins. Lee was my former teacher at the ranch. After two years, my sister, Virginia, graduated from high school and came to board with the Perkins also.

Shortly after that, my mother decided that she would like to come in from the ranch and bring my other two sisters, Pat and Phyllis to Alhambra and keep house there for all of us. We rented a nice house and furnished it. But my mother wasn't satisfied to just keep house. She went out and got a job in a defense plant and worked there until she moved back to the ranch.

Virginia was already in the Navy, and I was about to go into the Army. My uncle Aaron had stayed at the ranch during those years, but he owned a place in Huntington Beach and needed to be there part of the time. My mother felt that she needed to be at the ranch full-time, because when my uncle wasn't there, things just seemed to disappear from there.

So my mother and two younger sisters returned to the ranch. Mother sent Pat and Phyllis to school in 29 Palms and wrote a lot of letters to people who she thought might help my dad with his legal case.

Bill and Frances Keys in the kitchen of their Desert Queen Ranch about 1955.
Photo by and courtesy of Ross Carmichael.

THE MAIN HOUSE

Dad was single when he built the house. He had a fellow there, Ray Bolster, who helped him build the fireplace. Ray was a good designer, and he knew how to design a fireplace so it would draw good and put out the heat. My dad laid the rock.

The rock came from a rock cliff about a quarter mile beyond the south gate of the ranch. There was a little dim road that takes off to the right and goes back against the cliff there. That was our rock quarry and you see some slab looking granite half way or two thirds of the way up there. The rock came off in a sheets and my dad could cut it.

There used to be a tramway there with a little old mining car, a tram and windlass at the top. We could run that car up, and my dad would cut these slabs and lower them down and haul them in here. And that's what was used to build that fireplace.

They also put a barrel up on the chimney, because if the winds blew from the north and came over the rock, they would come right down and blow the smoke into the room. The barrel kind of eliminated that smoke problem.

I think that Ray helped him build the house too, but whether they ever drew up a set of plans or not, I really don't know. When Dad built the house it was one long room with the upstairs and a porch. The porch came around the south and part way on the west side. It was an all open porch.

After my dad got married and my mother came out, then he partitioned that bedroom off from the main room. Possibly that end of the house had been a kitchen to start with. They didn't need much in those days.

The little cabin that became a kitchen was later moved down and attached to the back of the last school house. I can't remember where that little cabin came from, but it became the kitchen. They moved that in from someplace and butted it up against the front porch and screened in a portion of the porch for a storage area for can goods and whatever. There was also a desert cooler in the back.

To get to the kitchen, you had to go through the door in the living room, across the screened porch, and then through another door leading into the little kitchen. It was a poor arrangement, but that's the way it was built on. Part of the living room was also used as a dining area until the old kitchen was taken off and the new one built around 1934.

9

Bill's rock wall increased the garden and orchard area and protected them from heavy rain runoff in the wash.

On the open porch sat the washing machine. It was run by one of the old type fly wheel engines - a Waterloo Boy. It was a stationary engine, and that's what ran the washing machine. In later years they got a genuine Maytag engine, a little two-cycle engine which had a foot starter on it. Then when they built the new kitchen and did away this setup here. The washing machine is now out in the yard.

I don't know where he got the lumber for the main house, but later extensions came from several buildings he tore down in Gold Park, the mining area just out of 29 Palms. There was a fellow by the name of Roach that had owned it, but he didn't keep up his assessment work. So Dad said, "Well, I'll file on it." So he did. He went over there and filed a mining claim and sent in proof of labor.

Roach came out and saw that my dad had taken it over and had a fit. He pulled a gun, and my dad shot the gun out of his hand. There was never anything done about that.

Roach said, "All right. I'll pull out."

So he got his personal belongings out of the place and pulled out, and Dad tore all of those cabins down. I think there were four or five of them, and a mill and a hoist. He moved all of that stuff to the ranch and that gave him some building material. The tent cabins were made out of these old mining buildings too. He had many stacks of lumber at the ranch.

There used to be big piles of stone that my dad had cut out in front of the house. He thought about rebuilding the whole house starting from the fireplace. He was going to build outside and then tear the inside down. This was probably about 1918, '19, or '20 - along in those years.

But he had too many other things to do trying to make a living and keeping up with his mines and their assessment work. It would take a couple months out of the year doing the assessment work on his mining claims and trying to take out a little ore too. Then he would have to make a mill run, work with his cattle, and haul hay for his horses. There were a lot of things to do, so he never did get around to do that rebuilding.

A few years later my mother wanted to have a new house built down there where the old garden was to the south. And my dad said, "Well, that's where we'll build the stone house."

This is vivid in my memory, because I hauled all of that rock out of the front yard on a little truck down there to where that building site was. I had a good excuse to drive it. So my dad said I could drive it and use a little gas if I hauled the rock. So that's what I did and hauled it all down there. In later years he hauled it back, and I think that most of that rock is in that rock wall along the wash.

The open porch went around the front, side, and back of the house. Later when there were more kids, Dad closed it in and made little bedrooms in there. And that made a little more room. I slept out there on just a little cot.

The room upstairs was where my grandmother stayed when she came out to visit. And when she wasn't here, visitors came, relatives, or close friends of the folks, and they would stay up there. Otherwise they would stay in one of the little cabins. But if they were someone close, they would stay upstairs.

In winter the only heating we had for the main part of the house was the fireplace. The old house of course had no insulation, so on cold winter nights we'd built a nice big fire in there and get up real close. We talked and played games and read the paper. My mother and dad both liked to read. We kids would play with whatever we had. We had some little toys and stuff like that. Along about eight o'clock was bedtime, so we went off to bed.

About an hour before we were ready to go to bed, we'd get ourselves a rock or a brick and put it in the coals to let it get good and hot. Then when we were ready to go to bed, we'd get it out of the coals and wrap it in newspaper and then a towel and put it in the end of the bed. Those rocks stayed warm most of the night, and that really helped!

Fireplace completed by Ray Bolster and Bill Keys in November 1917.

The dining table once belonged to Frances' mother, Lena Lawton. The children's high chair was made by Bill from the wooden sucker rod of a windmill.

The room upstairs was originally my grandmother's room, and it got cold up there in winter too. So Dad cut a hole in the ceiling and had a little trap door so that the heat from the fireplace could go upstairs.

We would take a bath in front of the fireplace or in the kitchen. Someone had to go out to the well and bring in a bucket of water that was warmed on the stove in the kitchen. Then we'd bring in the big wash tub, put a bucket of water on the sink top and sponge it on. That was the way you took a bath in the early days.

The bathroom on the back of the house was originally used by the mine workers at the Eldorado Mine. It had a shower and a dressing room section. Fred Vaile asked my dad if he wanted it, then to go over and get it. So he hauled it over and put it out here in front of the house and set it up. This was in the '30s.

Dad built a water heater with a place that you could build a wood fire under it to heat the water. You could get out there and build a big fire, and then get in and take a shower. Later he moved it over and attached it to the back of the house and added a tub. I don't really know if my folks ever got any use out of that or not.

In the early days we had a Victrola in the living room. It had a big wooden horn just like in the picture on RCA Victor records with the dog listening to it. That was our entertainment you might say. Then about 1929 the folks got a radio. But it was a chore to keep that radio in operation because you had to have at least three or four different kinds of batteries.

You had to have the main "A" battery which was an automobile battery, and you also had to have two "B" batteries, and you had to have a "C" battery. So it was a problem to keep a radio in operation. And you can imagine that there was a lot of static trying to tune in something.

I can remember that thing in a thunderstorm. My dad had a long antenna that went from the chimney of the house over to a post in the rock pile and out to one of the poplar trees. And during a thunderstorm that could pick up quite a bit of juice.

The radio was in a big cabinet which sat on a little table in the living room. The wire that came in from the antenna and one from the ground were about an inch apart for the hookup in the back of the radio. Every time the lightning would hit, the sparks would jump with a loud snap. Boy, that would get your attention!

Our last kitchen was considered fairly modern when it was built in the 30's. There were several people that helped. My uncle was there, Lee and Chet Perkins, and the Kilers. They moved out the old kitchen. They had the material there, and I think that in two or three days they had the new kitchen up. Chet Perkins was a sheet-metal man who did the flashing, roof-jack and stuff like that. They really went at it and put it up in a hurry.

There was a wood stove and a desert cooler reached through one of the windows on the outside. It was a screened-in cabinet with burlap sacks over the side with a water pan on the top. The burlap sacks came up and into the water pan and kind of siphoned the water and kept those burlap sacks wet all of the time. And the breeze blowing through it, cooled things. So we kept our food, our milk and butter and stuff like that in there.

I think that the oak kitchen table was my grandmother's. The original ranch table that we had to start with had gone to pot, and my grandmother gave my mother the present table. We brought it out from Huntington Beach, and we used it there in the dining room for years. It had about a dozen leaves that you could put in and stretch it out. After we put the new kitchen on, it was moved in there.

Frances' kitchen wood stove originally came from a hotel in Banning.

14

Willis and Virginia Keys and their homemade wagon,
April 1927.   Photo by Paul Allen.

The high chair was made by my dad from a sucker rod from a
windmill. He made it for Ellsworth when he was a baby and able
to get up to the table. They used to make sucker rods out of wood
and they ran from the pump up to the windmill head.

The Servel refrigerator was originally General Patton's
refrigerator. He had a camp way over in the Pallen Mountains and
had a little house built there. And that was his refrigerator. When
my dad located those gypsum claims there, he moved the propane
refrigerator over here and used it.

We mainly relied on the kerosene Aladdin lamps for light. In later
years we had a little Gottlieb generator that sat near the house
behind some rocks which kept the noise away from the house. think
we had the generators going in the early '40s. Then later on in late
'48 or early '49 we brought it up when it was cooler and set it up
in the barn.

My mother's piano was originally stored at my grandmother's in
Huntington Beach. And when she passed away, the folks brought the
piano up and put it in the living room. It was an old player piano
that you peddle to run the rolls through and it plays by itself.

In the early '30s the living room started to come down. The
timbers holding up the ceiling apparently were only lapped a little
bit and the weight from up above started to tip. So my dad said
we had to do something about that. So that's when he put these
posts in here in the living room. Before that they weren't in here.
Now the Park Service has put in extra supports for protection. So that
was the reason Dad put in these earlier posts. The place started to
go. It wasn't built quite right at first to support the room.

CRIB ADVENTURE

In 1925 when I was four years old, Virginia was a baby in a crib. It was partially screened and had four wheels on narrow tires. We kept it in the house most of the time, and all of us kids were brought up in that crib. My sister was just a baby and was maybe six months to a year old. She was still in the crib.

Well, my dad had gone to do the assessment work on some of the mines, and he had been gone for more than a week already. I was too small to go and stayed home with my mother who was worried. It's dangerous working around these old mines with hundred foot shafts with rickety ladders. So she was worried and thought something must have happened.

In those days there were no neighbors at all that you could go to in case of problems. Johnny Lang was probably down at his cabin, but sometimes it was a long time before we saw him. He'd go to town and sometimes spend a month there.

Dad was only supposed to be gone for four or five days. So my mother worried for a day or two and finally said, "We're going to go look for him."

I was just able to walk, so she put my sister in the crib and wheeled it out the door. The crib was about four feet long or a little longer and about thirty inches wide and probably thirty inches high with the wheels. The wheels were six inches in diameter with hard rubber tread but very narrow. So when you got to soft ground or sand, it was practically impossible to push it. But we did.

You can imagine us pushing that crib down that road. Those little narrow tires cut into the soft soil, and as we got down to the big wash crossing at the turnoff for the Cow Camp, I can remember my mother and I struggling to get that crib through the sand and up the rise on the other side.

After we got there, my mother said, " I don't think we can do it. We'll have to go back!"

So we did. We struggled back across the wash and back into the yard to the house.

My dad came in that evening or the next day. He had been delayed finishing up some of his work at the mine. But that trip with the crib has stayed in my memory all the years since.

WHERE'S THE WATER?

You know, I don't think this area is rich in water. There are springs up there in the Park, but they are few and far between.

In my younger days during the late '20s and early '30s, I saw a lot of people who came out there and homesteaded. There were more than you realize. If you go through there now, there is nothing. But there were quite a few people who homesteaded in what's now the Park.

Most all of them dug a well or at least tried. In those days a homesteader couldn't afford to hire a well rig to come in and drill a well. They dug it by hand. Some of it was rough - solid rock - and they had to blast it. They'd go a hundred feet, a hundred and fifty feet, and no water.

The Lost Horse Well was a little spring, and they called it Witch Spring at first. Later they built the well there. When the Lost Horse Mine started, they really wanted more water. So Ryan had a steam well rig brought out from Los Angeles, and they set up just about a half a mile above where the well is now up the canyon. Here they drilled a well 985 feet deep and no water - not a drop.

So they just gave up and used what they had. As far as I know, that was the deepest try for water in the area then. But most homesteaders had hand-dug wells 150 feet and that was as far as they could manage.

In places in the Park where there is so much rock and bedrock, you find places usually in a low wash area that have dampness well into the summer after everything else has dried out. I believe the early Indians that came through there probably would dig in areas like that and recover some water for a length of time. A basin in the rock underneath the sand or the dirt held the water there. That's why they would dig a ramp in there to get to it.

I drink a lot of water compared to what the old-timers did out here. They could go all day long in the hot sun and never take a drink of water. My dad was that way too. We would go over to his mining claims in the Pallen Mountains and my gosh, it was hot! And I'd take a canteen of water, and my dad wouldn't have to. He would never take a drink all day long.

Well, when my dad first came to what was later our ranch, I can remember him telling that the well there at the five-stamp mill was in operation and had a pump in it. I think that you can still see the remains of it where it has fallen in. It was right next to the arrastra. The well was right there. And that was one of their main water supplies for the mill at that time.

The McHaneys had dug that other well up where the windmill is now. And when my dad got there, it was in kind of poor shape. So he went in and cleaned it out and dug it down into the rock into the granite so it had a water basin. He put a pump on it, and it always produced good water.

Over the years we dug it down another ten feet I think and a tunnel out from it to get a water supply. The last six or eight feet are in granite and the water comes right in through that granite.

The well out in the middle of the orchard was also there probably dug by the McHaneys too. I believe that my dad used that when he first planted the orchard and raised a little garden there. It was about eighteen feet deep.

My dad was plowing one day with a team of mules and one of the mules fell in it accidentally. He had a heck of a time getting the mule out. He rigged up a chain block and got it out. Then my dad covered the well over.

There was a tunnel underneath that went from the well all the way out across the wash. He went down there and drilled a hole with a post hole digger into that tunnel and put an 8 or 10 inch casing in it. Then he put a little gasoline engine-driven pump on that.

That pump was right out toward the wash. It was just inside the old fence which would be in the orchard area from the wash now maybe eight or ten feet or more. It was probably about 40 feet between the well and where the casing was. And from the casing it still ran all the way across the wash underneath. My dad used that well a good deal for watering in the garden area and probably for drinking water too.

The well dug over by the arrastra was probably dug when the McHaneys put in the mill. Just across the wash from that was one of the old original wells which I think George Meyers dug. There are some poplar trees or cottonwoods growing out of it now.

Even before that down there below the old garden right alongside the road, there was another old well. And I think that might have been a George Meyers well also. I remember Bill McHaney telling my dad about those old wells there. And some of them were there when Bill McHaney came in there. And I think that George Meyers dug those.

When Dad first came there wasn't any lake or dams there at the ranch. After a rainstorm, little ponds would form up there in that channel. In the winter months when we had a little rain, water would stay in that area for maybe six months at a time. Moisture coming down off the rocks would seep out through the rocks at the bottom and flow down the creek bed. It would keep trickling down and run for quite a while after the rain or after the winter. I think that gave Dad the idea to first build the dam here.

His first dam was a cement one which he built in 1914 or '15. It was the north dam, the furthest one up the wash. He had a fellow by the name of Thompson help him. Now whether Thompson was learned in that sort of thing or not, I don't know, but he and my dad did a pretty good job. It was shaped like a dam should be to give it strength to hold the water back.

After he built it, the lake formed behind it, and as the water level came up, he found it was going to run out where the south dam is today. So then he built the dirt levee in there to hold the water.

The earthen dam went out the first time in the summer of 1931 after a heavy rainstorm. We had a bunch of goats who walked over it to get to the lake and they wore a little "V" in it weakening the levee. We rebuilt it using just burros and scrapers and sandbags and stuff like that. But it kept going out every time we had a big rain or cloudburst because the water would seep underneath. After it went out the second or third time, Dad said that we would have to build a concrete dam. This was about 1936.

The Tucker boys were with us then, and they spent probably a month digging a huge trench across the opening down from where the levee had been. You can still see the remains of it right across the whole area. And we were going to build a dam in there.

Then my dad decided that was just too long, so he went up closer towards where the old levee had been. The dam started out with rock and cement. We hauled big rocks in and cement and built it up quite a ways. It went pretty slowly. It's four feet wide at the bottom or better, and Dad was laying it up piece by piece. He would cement a rock wall in front, then fill it back with concrete, and rock in the center.

It was several years in the building. He worked on it in his spare time. I was just a kid when he started that from the bottom. I helped him on it, and later on, my sisters helped him too. My uncles also worked on it.

You can see that at the bottom where he laid the rock, and then he got up so far and decided that he would form it inside and out from there and pour the concrete in. So the first part was done in the middle '30s, then it was terminated for quite awhile. Later on in the early '50s he went ahead with forming and pouring the higher walls.

He used a little bit of everything in that cement - any iron, steel or stuff like that, even cable for reenforcing here and there. As he built it up higher, and added to the main dam, the lake ran over back of the old schoolhouse. So he had to build a concrete dam up there in back of that cabin.

So the north dam is the first one followed by the south dam to replace the earthen levee, and the one behind the school house was the third. In between the additions and finishing the one where the levee was, he had gone over to Barker Dam, raised it, and refaced it. The last addition to it was finished in 1950.

I remember we moved that cable tram from the Desert Queen Mine down to Barker Dam and started on the last extension of it. I helped him on that project, and my sister and mother helped him too. Phyllis ran the hoist to get the materials up to him at the dam. From there he moved his cable tram over to the Cow Camp and set up, and she helped him there too.

After he finished Cow Camp, he pulled down the cable unit and concentrated more right around the ranch in the area back of the house - the rock work, cement work, and the new dam. He intended to dam the overflow coming out of the lake over the north dam and make a pretty good size pond in there. But he never finished it.

Some of his dams don't have the shape of the original one, the north dam, that he built at the ranch. In some like Barker Dam and Cow Camp he included existing rocks that were in the way into their design. Those he worked into the dam. A professional dam builder would have probably blasted those out, but my dad included those rocks in the dam itself. He liked to build things, work with rock, and work with his hands all of his life.

Bolster family members enjoy a 1920s swim in the Keys Lake.
Photo courtesy of Harriet Austin McClowry and Walter Bolster.

When the McHaneys and George Meyers lived at the ranch, they didn't have the lake or dams, but they did quite a bit of well work. The only dam the McHaneys did was the cement and rock cattle tank over at the Cow Camp.

Right near the house was the well where we got most of our household water. I don't know what year it was that my dad dug that well, but he dug that with a post hole digger. You might wonder how anyone could dig a 20 to 30 foot well with a post hole digger.

Well, you just put extensions on it, hoist it out, and dump it, then lower it back in. The regular handles unscrew. It's 3/4" pipe and you can make your own extensions with more pipe. The further you go down, you just put on another extension, then screw the handle back on.

The post hole digger was one of the rotary types, like an auger. It wasn't like the two-handle shovel-type that people use today. You wouldn't get very far with that. It was a rotary.

After he drilled it out, he put a metal casing in. He didn't have to put the casing in as he drilled, because it was all clay down through there and didn't cave in.

That well provided the water for some of my chores. One when I got old enough was to keep the water bucket filled in the kitchen. We had a pail and about two or three times a day I had to see that it was filled.

The old cooler that we kept milk and other perishable food in used to sit out on the porch. It was a wooden framework with burlap and screen over it with a little pan on the top. And I used to have to keep that filled too. Two or three times a day it would have to be filled with water - water that came from that well close to the house.

In later years my uncle Aaron took the casing out of the well and dug it out. He got in there, dug it out by hand, and bricked it up. Then he put the arched framework over it, and added a bucket and rope to use to pull the water out.

There also used to be a big stand of bamboo next to that well. My mother liked bamboo and took care of it, and kept it watered. When we got fish in the lake, I'd cut a pole every once in awhile and make myself a fishing pole.

That well was used for years and years, and as long as I can remember it was our main drinking water supply for the household.   When Dad built the first cement dam, he put provisions in for an irrigation pipeline. And as soon as he got that done, he ran that pipeline down into the orchard area.

We also had fish in the lake brought in by the Kilers probably in the middle to late '20s, because I think there were fish in the lake when the dirt levee broke. The Kilers used to try and come out quite often and would spend two or three weeks here in the spring and summer.

Finally Warren Kiler said, "I'm going to bring some fish out." So this time when they came out, he had two ten-gallon milk cans with black bass. Then later we brought out some blue gill and put them in. And they did well.

 I could catch some blue gill once in awhile, but I wasn't much good at catching black bass. My uncle Aaron had the patience. He would go up there and bring back a batch of them.

My dad tried some catfish once, but they didn't do too well. He used to keep records of rain for the Department of the Interior. They brought up a rain gauge and set it up there on a concrete slab.

We went swimming in the lake in the summer. When I was young, the old dirt levee was here and that was where we all went swimming. There was a gradual slope up to the top of the levee, and that was the nicest place to swim. It was fairly wide back there.

Later after the dam was there, we had a raft. It was eight feet wide by twelve feet long. We'd pole it out and then swim off of it. That was about all we had in the boating line.

In the early days the lake never dried up. We had enough rain every year usually anywhere from six to eight inches. But after the levee washed out the second time, it lowered the capacity considerably. This was also when we started to have some dry weather in the early to mid '30s.

So the lake completely dried up about '34 or '35, and we lost all of the fish. The wells got low, but they never dried up completely. We hauled water from Quail Springs for the Wall Street Mill, because that well would get low and the mill took quite a bit of water.

Dad appreciated the conservation of water and its value as a resource. Where you have water, you could do something. You could run cattle, raise some crops, or run a stamp mill. None of these are possible without it. So he taught us the value of having water and the wisdom of never wasting it.

A rock wall, framework, and bucket were added to the ranch's household well by Willis' uncle, Aaron Lawton, in 1944.

RAIN AND SNOW

When I was young, you could count on several thunder showers during the summer and probably a foot of snow during the winter which made for good water and good cattle range too. During those years the lake never went dry. There was always water there.

Then in the late '20s we had about two years of drought with hardly any rain. The Barker-Shay outfit were running cattle up here then, and when the water holes went dry, they didn't come to get their cattle. They just left them. There wasn't any feed and hundreds of them starved to death There were cattle carcasses all over.

So my dad called the Humaine Society and tried to get them to do something. They came out and looked around, but since that cattle outfit was owned by members of the County Sheriff's family, nothing was done.

So the cattle continue to starve. There was very little water any place. Then the rains came back a little bit.

In the early '30s or a little later after the dam had washed out here at the ranch, there wasn't as much water in the lake. Then it went dry, and we lost our fish.

There were couple of years when it was pretty dry, and then we would have pretty good rains again off and on. Now I look back at the difference in the vegetation between then and now and see that a lot of the old vegetation is gone.

These dry washes through here all used to be lined with wash willow. There were big ones, and some were maybe twelve or fourteen feet high.

Usually it wasn't cold long enough for the lake to freeze in the winter, but it did freeze around the edges. That winter of '48-49 it did freeze and the ice was about two feet thick. It was probably the only year that I remember when we could actually ice skate on it.

We used to ice skate on a little pond back behind the house that was caused by the run-off from the north dam. It would freeze over so that we could skate on it. Actually we would just skid along on our shoes. We didn't have skates.

ORCHARD and GARDENS

Originally the orchard was just fruit trees. Later on he planted a garden here too. Over on the south side there used to be a big almond tree. It was huge and spread way out. We used it as a picnic ground in the summer time underneath that shade.

Originally there was a lot of clay in the orchard and garden and the soil was not that good. So my dad hauled in dirt and stuff out of the lake bed to built it up.

When he originally planted the pear trees about 1914, he would drill holes with a auger and then put three or four sticks of dynamite down them and blow a hole. This would loosen the ground enough so that these trees could take root and grow. But the roots fill out the hole and finally they couldn't get out through that clay. So it wasn't too good. The trees used to get root bound.

Dad really had to work to make the soil better. We had horses and cows and all of their manure went on this. And then in later years he went up to the lake and cleaned out the silt from there and hauled it down and put it here to help the soil. But originally the soil was just adobe.

Dad had the orchard pretty well lined up and spaced evenly. There were several kinds of pears, several kinds of apples, some pears, peaches, and plums. And for awhile they all did real well.

When one of the trees died, he replaced it. He used to get a lot of his trees from a nursery in Yucaipa.

As I got older, I remember using an auger when we'd replant some of those trees that had died. One of my chores would be to get down there to drill those holes so we could put some powder in there and blast.

In the mid '20s he also had a little space down in the orchard for the vegetable garden. We grew everything in the way of vegetables - beans, corn, cucumbers, pumpkins, radishes, squash, and tomatoes. We also raised a lot of melons - watermelons, cantaloupes, casaba, and honey dew. We had just about any kind of vegetables imaginable.

He usually raised a pretty good crop of corn - both sweet and field corn. The white corn was called Country Gentleman and the yellow was Golden Bantam. They were good eating.

The field corn was for the animals. They were big ears - more than a foot long - which had pretty coarse kernels. Some of those stalks would get six to eight feet high.

My favorite tomato was a pink one. I wish I could remember the name of it and find it again, because it would be a treasure now.

It had a very thin skin which was probably the reason it wasn't continued, because it wasn't of commercial value for shipping. It was a pink color and kind of flat but wide shape. Boy, it had a great flavor all it's own.

My dad trapped quite a number of rodents in our garden. He had fine mesh chicken wire mostly around the garden at one time. But it was a constant chore because the squirrels would dig under and others would climb over. So he was after them all the time. And we had gophers too.

We had cats from time to time, but they couldn't take care of them all especially when we had the garden down in the lower place. They had it fenced in the same way with chicken wire and other wire. I can remember him setting traps around the outside of the fence all of the time trying to catch some of the animals coming into the garden. In spite of the rodent problems, we raised a very nice garden down there and up here too.

Usually my dad would be down in the garden part of the day working. And in the evening, just about sundown because it would be hot and it would cool down a little then, the whole family would go down there and pull weeds. That was a big thing. There were always a lot of weeds to pull. It was kind of a daily routine to go to the garden and work for a couple of hours.

And then when we got ready to come home, my dad would pick a bunch of vegetables and watermelons and load the whole back end - the back seat and floor boards. It was just a big load of stuff.

Although my mother canned a lot of the vegetables, we also gave away plenty.

The orchard and garden were watered from the dam and the windmill. He pumped most of the water out of the well, and then we'd use the lake there too.

The pipeline came from the north dam, the one furthest up the wash. It came up into the orchard by a different route than it does now. It came down along the bank, then across the creek near where the unfinished concrete dam is, then up into the orchard by the corner of the house, and then out into the garden where it was distributed by movable pipes.

Later on my dad changed the lines and put in a 2" pipeline right to the southeast end of the machine shop. When he decided to make that garden down below, he ran a another 2" pipeline from the machine shop all the way down there. He put it in a ditch two feet deep. That was quite a chore because he dug it all by hand. And that 2" line was what watered the garden. That has since been taken out.

Water from the lake was used to irrigate the garden below the house. The 5-stamp mill originally operated by the McHaney brother is in the background of this 1931 picture.

Two visitors with Bill Keys (on right) in the garden below the house about 1933 or '34.

The area around the old 5-stamp mill was fenced in and used to be planted in grain. I believe it once was planted with alfalfa. We raised alfalfa and fed it to the horses and animals. We also used to chop it up for the chickens so they had greenery. We'd go out every morning and cut an armload and take it out to the cows or horses.

The pipe in the rocks there was part of an old shed that was in use at one time. That pipe was part of the framework.

Dad built that wall along the wash to protect the orchard area from washing out. In earlier years, we'd get some good floods down through there.

That canyon goes quite a long ways back of the dam. And there are little streams that come in, and it's mostly rock. And the rock sheds the water and comes down all of these little streams. We used to get these cloudbursts during summer thunderstorms. They would come up and it would rain for about an hour. That was the extent of it.

That would put down an inch or an inch and a half of rain. When it hit those rocks, all of it ran off, and it all funneled down right into that area where the dam is. Of course, when the lake filled up and went over the dam, it went down the wash back of the house.

I've seen it running in that wash five or six feet deep - really a torrent of water. I've also seen the water come half way up to the house.

We were always thankful for a thundershower, because that was when there would be a lot of nitrogen in the rain from the lightning. It would make the garden green up right away and make it grow a lot. Dad saved rainwater for batteries because it was fairly pure. Mother used to put her dish pan out and get it full for washing hair. The water was really soft.

Originally Dad had poplar and cottonwood trees planted along the edge of the wash and that kept the erosion back. Then he felt that the trees were taking too much of the water and the nutrients from his orchard. By then he was raising a little alfalfa down in that area too. So he cut all of those big beautiful poplar trees down.

Then he decided he probably ought to have something there to keep the erosion back, because the trees were gone and the water would start cutting in. That's when he decided on the wall. He needed some rock work to do, so that was a good project.

When he built the wall he extended it out into the wash a little further than it had been, and he gained a little ground in his orchard area. He started to do it in '39 or '40, along in there, but most of it was done after he came back from prison.

My dad liked to work with rock. He was building another dam behind the house to catch the overflow of water from the north dam to form another lake there. That area there by this dam was my dad's rock workshop. He rigged a hoisting arrangement there so he could move the rocks around to place in his dam.

There was a lot of rock in that area where that crane boom is. All kinds of boulders were in there, and he cut them all up right there on the ground. That was a good supply for him, and he could just reach out with a cable on the end of that crane - he had a hand winch on it - and he could lift them up and set them in place on his wall.

He did everything by hand with a hammer, a chisel, and a drill. You can still see the little drill holes along which those rocks had been cracked. He drilled holes about 3" or 4" deep in line where he wanted to break the rock. Then he would drive steel wedges into the holes, and hit them one at a time down the line. Then he would go back again until he had created quite a pressure that would crack the rock.

Originally the hoist was closer to the wall, and after he had completed that section, he moved it to work on the next one.

He quarried a lot of rock out there beyond the ranch's south gate. Wherever he found acceptable rock, he'd take a crow bar and put up a ramp or something and work it into that trailer that he pulled with his jeep. Chunks of rock too big were wrapped with chains and dragged back to the ranch to work on there.

Over the years Dad made many grave stones and stone markers. Some were for Johnny Lang, Bill McHaney, and Worth Bagley. He made the headstones for my mother's grave and those of my brothers. He did this all by hand with a hammer and chisel.

I carved my dad's marker on a trip back to the ranch on Easter Sunday 1978.

SNAKE ENCOUNTERS

Snakes used to come into the yard quite often every summer, but none of us ever got bitten even though there were some mighty close calls. And from the time we could walk, we were always cautioned to look out for snakes and to be careful. And we were.

There were good snakes there too - gopher snakes, red racers, and water snakes. And we always left them alone. Actually as I remember, we didn't worry about rattlesnakes too much. When we found one, we got rid of it.

We just watched out when we were walking in the summertime. Rattlesnakes in the daytime especially like to get in a little shady spot in the grass, so we looked when we got around places like that. We were really careful when we moved anything.

I remember one time we all had walked down to the garden to do the weeding, and we were coming back at dusk. I was barefooted, and I had forgotten something. As I ran back in the low light, I saw this thing lying there and thought it was a stick. As I stepped on it, I knew it wasn't a stick. It was a rattlesnake, and I just kept on going!

When I was about three years old, my dad was working on the trees in the orchard behind the house, and he had me back there. I was sitting on a rock, and our little dog was sitting beside me. Well, the dog sat there for awhile, and pretty soon, he jumped from the rock that we were on to another one a few feet away.

As the dog jumped over, there was a snake in between and bit it in the chest. The dog let out a yelp, took off, and crawled under the house.

Even ranch dogs and cats could be bitten by rattlesnakes.

Well my dad grabbed me, took me into the house, and killed the snake. Then he tried to get that dog out. The dog had become paralyzed from the poison and couldn't move. He was trying to crawl out but couldn't. So Dad took a long pole with a hook on the end and fished the dog out from under the house.

He didn't have a lot to work with, but he did have a big bottle of ammonia there in the house. The dog was unconscious , so he cut the wound open, let it bleed a little bit, then turned that bottle of ammonia upside down on the wound.

Dad later told me that he could see green stuff coming out of the wound as the poison reacted to the ammonia. He held it there for a long time before he took it off and bandaged it.

Well, the dog recovered but lost a lot of its hair and part of its hide around the wound from the ammonia because it was so strong.

We used to keep the horses up in the back pasture, and sometimes I used to go after them. One day as I was walking where the trail narrows with rocks on both sides, there was a snake under a rock shelf. As I walked by, I might have kicked some dirt on him and woke him up. Well, he struck, but just hit my pant's cuff.

Another time my dad and I were out surveying the property line over their by Barker Dam. We were way up in the rocks, and I was pulling the tape. My dad had the other end, and I jumped down off the rock underneath a little oak tree. There was a bunch of leaves there, and apparently there was a snake in there because it buzzed when I stepped down. Well, I kept on going!

Another time we were up at the Desert Queen Mine over on the hill surveying the line there. There were three of us - my dad, myself, and Oran Booth. My dad had his steel tape and was measuring.

Well, it came time to rest, and we stopped by an old shaft. My dad set his tape on a rock and it got knocked down the shaft.

The next day my dad said that we had to go back and get that tape reel. So we took a rope and he and Booth held it while I climbed down into the shaft. The shaft went down and then there was a step off and it went down again. I made it down to the bottom where there was a lot of brush, and before I hit bottom, I could see the tape reel.

Just as my foot hit bottom, I could hear this buzz. He was in the corner. It was almost dark down there because the shaft was offset a little bit, but I could see the reel and the brush in the corner. So I jumped back on the rope, edged over, and got the reel. Then I hollered for Dad and Booth to haul me up. I was down about forty feet I guess.

After I went away to high school in Ontario, I came back to the ranch in the summertime, I would sleep in that little cabin up in the canyon. That was my bunk up there. In the morning I'd walk down to the house, have breakfast, and then go riding after the cattle, or whatever I was going to do that day. At night after I had supper, I'd walk the hundred yards up from the house in the dark to the cabin to go to bed.

I didn't have a lantern or a light, but I did have a kerosene lamp in the cabin but I never used it. I wouldn't even bother lighting it. I'd just sit down on the bunk, take my boots off, and crawl into the sack. And that was it.

So one night I walked in, and when I sat down on the bunk, I heard a buzz. I knew what it was immediately. I'd left the door open that day and a rattlesnake had come in. It was under the bed. I guess I moved pretty fast and got out of there!

Then I went back in, got that kerosene lamp going, and got him out. It kind of shook me up a little bit, but that was to be expected if you had gotten a little careless and left the door open.

Remains of the family beehives.

BEES

My dad could trail bees by standing at a watering place and watch a bee leave. He would watch the bee go as far as he could see him, then note the direction and pick out a land mark. Dad would walk to that mark, watch for another bee, and so on until he found the hive.

Other times he would find a swarm of bees in a tree. He would cut a section out of the tree, get in there, and try to find the queen bee. The other bees would be around her and take care of her. So he'd saw into the tree, grab a bunch of bees, put them in a wooden hive and put the lid on. If you put a little bit of food in that new hive with the queen, the other bees would usually move in and build up the hive.

Once in a while a swarm would take off, and you'd see a whole big cloud of them going out through the air. They'd usually light on a tree limb somewhere and cluster in a big bunch with the queen in the middle. So as soon as the swarm got settled on the limb, we'd go under them with a hive, tap the limb, and they'd all fall in the box. Then he'd put the lid and put them where we wanted them. In the spring the bees would make some real good honey from the sage and other wildflowers.

Originally, he built his own hives, and later on he bought regular ready-made ones. I guess at one time, Dad must have had a couple of dozen wild bee hives that he had gathered.

I would often go with my dad to rob the hives. In earlier times I would stand quite a ways back from where Dad was working. Then I would get stung during the process, but Dad very seldom did. I couldn't understand this at first. Gradually I got braver and got up with Dad at the hive. After that, I seldom got stung.

I believe that the bees close to the hive were so busy trying to get their brood and honey out that they didn't have time to sting. While those further out would attack anybody who might be approaching the hive.

DEER AND BIGHORN SHEEP

My dad never went deer hunting. He'd rather see them and the mountain sheep than to hunt them.

He said, "As long as we've got beef, that's what we'll eat. I like beef better. There's nothing wrong with venison, but I like to see the live ones moving around." So far as I can remember, we never had a piece of venison.

I remember my dad saying years ago that when he first came out here, there were some deer in the area. But when I was growing up, I never saw a deer out here. The deer had kind of disappeared out here until the late '30s. We rarely saw any, but in the late '30s, they started to come back again. By the '40s there were quite a few deer around. They had multiplied a little bit and moved in again.

But there were mountain sheep, and we would see them quite often. They used to come and stand on the rocks once in a while. You'd see a big old buck standing up there especially if the mill was running or some activity or noise was going on at the ranch. They'd sit there and stare for a half an hour. You know, watching.

And when we used to run the old two-stamp mill over at the Wall Street millsite, they used to come and stand up there on the rocks. There might be four or five of them at a time and they'd watch. They were beautiful.

"Bill," the ranch's resident bighorn sheep, was a common site during the spring and summer of 1996 and 1997. Photo courtesy of and by Harmon and Nelda King.

SOLVING MEDICAL PROBLEMS

In years back because there weren't any neighbors around, my dad had to figure out how to do something without any help. Once in awhile there would be a homesteader in the area, but most of them only lasted a year or two, so we learned to do everything by ourselves.

It was the same way about getting hurt or being sick. You just had to be careful, because the nearest doctor was miles and miles away.

My parents had a set of medical encyclopedias and other medical books. And they kept up on that stuff pretty well, because they knew they had to take care of us.

I remember one time when my dad dropped a hand drill on his foot. It had about a three-eighths drill in the bit, and it went right through his foot. My mother put iodine on it or whatever she used in those days, but there was no way to clean it out completely. As a result, it got infected and swelled up.

It was getting worse, so they finally resorted to a bag of fresh cow manure which they put on it and kept it there. That drew out the poison apparently, because it healed.

I remember doing that also for one of the burros that ran a nail into the tender part of his hoof. Dad tried everything, and it still got infected. Finally, he used a bag of fresh manure and tied it around his leg. This worked and drew the poison out.

We kids had the usual ailments. We'd have colds, maybe a touch of pneumonia and scarlet fever once, and we were all treated at home. Most of our hurts were cuts and bruises and scrapes from falling on the rocks. But there wasn't a whole lot. My mother had a bottle of iodine, and bottles of ammonia, vinegar, and witch hazel. I'm not sure whether she had peroxide or not. I know when we got ant bites, she'd take that witch hazel and that would take the sting out.

About the only time anyone went to the hospital was when my mother was having another baby at Loma Linda. She'd go and spend a week or so there, and when she came back, she would bring a lot of information from there. My mother went pretty well by what the Seventh Day Adventists had to offer. She would also clip out articles, and if she thought they were really important, she would tack them up on the wall in the kitchen. These were home remedies and things like that.

My folks were always worried about us kids getting bitten by snakes, but in those days there wasn't a whole lot in the way of remedies. My dad always had a bottle of pomanganate of potash. That was suppose to be a remedy. You were suppose to cut the wound open, let it bleed, put this pomanganate of potash on it and that would draw the poison out.

In later years they said that pomanganate of potash was probably a worst poison than the snake venom in a lot of cases. So they don't use it anymore, but we had it. Probably we were lucky we didn't have to use it.

Horses are pretty well able to take care of snakebites, and cattle too. I had a horse that got bitten once or twice and it swelled up. And I think we treated it with ammonia and in a few days it was back in shape again.

My dad had back trouble - probably a ruptured disc - which gave him a lot of problems. He found out if he could stretch his back a little, it would give him relief.

So he built a traction machine and put it out behind the house under the trees. He would come out there, lay down on this bench, and put his arms behind the uprights.

He had a couple of old cut-off boots that he would put on and hook this cable to them. Then he would have somebody turn the crank, and he would get a little stretch there. He'd relax and lay there for a half hour or hour and get some relief.

Now they have apparatus where you can turn upside down and hang by your feet. This was the same idea but many years earlier.

The back stretcher helped alleviate Bill Keys' periodic back problems.

NO TIME FOR PLAY

Well, when you look at things as people do now, they think that it must have been pretty rough living and pretty lonely out here at the ranch. But you know, when we lived here we never even considered it. It was just normal living.

"How did you spend your time?" people ask.

Well, we really didn't have any spare time. Mostly in our younger years as kids, we played outside climbing rocks. Our toys were usually pieces of iron or wood or something that we imagined to be something else,

As my brother and sisters and I were growing up, our parents gave us chores to do around the ranch. And it seemed as we got older the amount of chores increased. Then there was no time for play. There were too many chores to do. I'd get up in the morning before sun up with my dad.

I think I started by keeping the water bucket full in the kitchen. And then a little later I had to keep the cooler pan full for the desert cooler. Eventually I assumed the chores of bringing in wood for the kitchen, feeding the chickens, and gathering the eggs.

Of course we had several milk goats, and they had to be milked every day along with the one or two cows that we had. And they had to be fed too along with the horses.

During the summer we helped our parents can the fruit and vegetables from the orchard and gardens. It was quite a chore to can and we all helped.

We had a couple of long tables outside where we would sit and peel the fruit, cut it up, take the seeds out, and get it ready. My mother and dad would then pack it in Mason jars and pour the juice over it. Some of vegetables took a little salt, while the fruit of course took a little sugar. Then they'd screw the lids on loosely, and take them over to the fire where the big copper wash boiler was half-full of water.

They'd put maybe fifteen to eighteen jars in that wash boiler, get the fire stoked up and get it boiling. My mother would time it. It had to boil for four hours. Then we could take the jars out, let them cool, tighten up the lids, and put them away. Then they put another batch of jars in.

My job was usually to keep the wood fire going under the wash boiler. I had to get the wood and to be sure that the water didn't stop boiling. While my mother always used to can the vegetables and fruit in the old quart Mason jars, she used the small jars for things like jams.

About every two weeks we'd have a wash day, and that was also one of our chores. We had this old gasoline-driven Maytag washing machine, and it usually took a couple of hours in the morning to get things started.

We had to build a fire under the wash tub to heat the water, fill the tubs and washing machine with warm-hot water, then get the gasoline engine started.

We'd all get out and help with that. A couple of us would be at the washing machine putting the clothes in, then taking them out, running them through the wringer and hanging them on the line to dry. So it took about a full day's work to do the collection of two week's wash.

As a kid, I used to get into things and scatter things around. And about once a month, we'd have a clean up day. They'd get all us kids out and clean everything up and put things back where they belonged.

So our leisure time was pretty scarce from daylight until dusk. There always seemed to be a chore to do, and that had to be done first!

The Model 30 Maytag wringer-washer's manufacture date of July 1929 was confirmed by Walter Cripps, Regional Administrator of Maytag's Los Angeles office. He took one of the Park Service's weekend ranch tours in 1994 and researched its serial number.

CATTLE TROUBLES AND A SHOOTING

You once asked if our family considered ourselves cattlemen or miners. Well, that's kind of hard to answer. During the days when I was up here, you really didn't give that much thought. This was our home, and we did a little mining and a little milling and ran a few cattle. We really didn't consider ourselves too much as cattle people and probably not as big time miners either. We were just doing what we could. As far as I was concerned, there were no sentiments either way. Anything that came along that we could do and make a living was the way we did it.

Barker and Shay, who were partners at one time, were strictly cattle people, while other Shays were sheriffs of the county. They didn't dabble in mining. Some of the others like the Langs came up here as cattle people and then went into mining. The McHaney brothers were the same way. They came up here with cattle, developed the ranch, then let the cattle business go and went into mining. They made the ranch their millsite. There were other people who ran cattle up in here, but they were small operations.

When Barker and Shay ran cattle up here, Cow Camp was their headquarters. They were not infringing on the ranch, because at that time the ranch millsite and the Desert Queen Mine were owned by Morgan, the McHaney's successors. Because there was water at Cow Camp, Barker and Shay established a little camp there and made it their headquarters for this part of the county. They had quite a little establishment there. There were a couple of cabins, a blacksmith shop and several corrals. They also had a well and a rock tank to catch water.

So after Morgan died, Dad filed on the millsite. About 1917 he filed for a 160-acre homestead that included the Cow Camp. And that's when the trouble really started. The Barker and Shay outfit were still using it, and that went against their grain when my dad homesteaded the land, fenced in that area, and cut them off. And that's when the trouble really started.

They decided to go in and take those buildings. So they sent in one of their cowboys who was a friend of my dad's to visit him and spend the afternoon. My dad was friendly to him, and they talked, and he invited him to stay for dinner. I don't know if the cowboy stayed all night or not, but he stayed late.

In the meantime the outfit had a crew ready with two or three wagons to go into Cow Camp to dismantle the little cabins and the blacksmith shop and its equipment, and out they went. By the next morning they were gone, and the only thing left was the rock chimney.

By law when my dad homesteaded the property, he was entitled to everything on it. So as soon as he discovered the buildings were gone, he went into San Bernardino and filed suit. It came to trial, and my dad won a $500 judgment.

When nothing came, he went in to find out about it. Well, it had never been entered in the court record, so there was no record of the judgment. As a result he couldn't collect. He didn't have the money to fight it any further, so he let it go.

Dad used the Cow Camp to satisfy the requirement for proving up on the homestead. The government required that you put in a certain number of acres of vegetation. It was supposedly a profitable crop which would be grain. So my dad cleared that area out, plowed it, and also another little field out away from there, leveled it out, and planted barley and wheat. I think that he planted some rye there one year too. It was beautiful. That stuff was six or seven feet tall. And that met his requirement for getting a patent on the property. He also grazed cattle in there too.

Above Cow Camp there is a little narrow valley. There was good feed in there because of the rocks and the water seepage. It had dampness way after any other area out here had dried out in the summer. So it had a little bit of pasture, and my dad would take the horses up there occasionally and let them graze.

When Barker and Shay were here, there were also other people running cattle out here. The Cram brothers who operated out of the East Highland area of San Bernardino were running cattle in the Park's Cottonwood area as well as other Low Desert areas. They had a lot of cattle over in there, but in the late '20s they pulled

Rock tank at Cow Camp, built by George Meyers, later held water for Barker-Shay cattle in the early 1900s.

Barker-Shay cowboys ran cattle during the 1920s.

their cattle out and left a few behind that they couldn't get. Some were pretty wild and had gone up in the hills.

So one of the Crams knew my dad, and he said, "If you can catch 'em, you can have them."

Well, my dad found one - a big, black steer. He found him over on the Pleasant Valley slopes, and he brought him back to the ranch. Dad put him in the pasture up above the lake where he could get plenty of feed. We had it fenced, and he'd be safe there.

In the meantime Shay had bought most of Cram's stock including the ones still wandering loose. Somehow Shay found out that my Dad had the black steer and made a deal with Cram for it. Neither of them let my dad know about the sale.

So early one morning way before sun up, a San Bernardino County deputy named Larson and four or five other deputies drove in. Dad was already up and the rest of us were just getting up. He met them out front where the old kitchen was.

"We came to get the black steer," they told my dad.

"You can't have that black steer. That's mine."

"No," Larson said, "That belongs to Shay. Shay bought it from Cram."

"Well, Mr. Cram gave it to me," my dad said."

"No. Shay bought all of that stray stock that was Cram's and that steer is part of it. We've got a bill of sale for them right here."

"Well, you're not going to take him."

Dad had a shotgun back in the house, and he started to go back inside to get it when a couple of the deputies grabbed and

A rock chimney is all that's left of the Barker-Shay buildings at the Cow Camp area of the ranch.

handcuffed him. And then they hollered down the road to Homer Urton, one of the Shay cowboys, who was on horseback. He knew right were the steer was and went up and ran it out. Urton had probably been around there before and had spotted it. Dad told him then that he didn't ever want to see him on his property again.

The deputies held Dad until Urton had left with the steer, then they turned him loose. There wasn't much my dad could do because they had this bill of sale for the steer.

I think my dad had had a run-in with Urton sometime before at roundup time. Dad used to go to these roundups because the Shay cowboys would gather his cattle right in with theirs. He didn't have many - maybe eight or ten head at a time. So he always made it a point of being at the roundup when they were gathering cattle to make sure that he got his cattle back. And I believe that he had had words of some kind with Urton then, had tackled him, and knocked him down. Urton had supposedly said, "I'll be after you.".

Anyway, during the next year despite Dad's warning, Urton had been coming through our place every once in awhile. There was a road around it too. He could go around through the Lost Horse Valley and through the little pass by the Whitlow's west of us. If you were riding horseback, it was just as close to come through that way. But Urton would make a point of riding through our ranch. So my dad posted a sign at the south gate:

> "Homer Urton, you are hereby notified not to
> set foot on or in any wise cross or enter upon
> my land or property. Hereafter as you will be
> considered a trespasser and treated as such."

So it wasn't long after Dad had posted that sign that Urton came through in a car with two women and another fellow. When he saw the sign, he tore it up and threw it in the road and went on through the gate.

My dad was going over to the Desert Queen Mine later on that morning and found the sign lying in the road. He knew right away who it was that had gone through. Dad was driving that old iron wheel truck, and he followed the car tracks down the road and saw that they turned into the Barker Dam area. So he went on to the mine and worked there for several hours.

Later that afternoon he heard a car drive up over the little hump where there was a circle turnaround. My dad was parked way out on the point, and he put down his tools, got his .44 rifle, and started back up the road. Then Urton came over the hill and stood there a moment looking around and didn't see my dad.

So my dad hollered to him, "Is that you, Urton?"

"That's me!" he answered.

Then Urton reached for the gun that he had in his rear pocket and got it out when my dad shot him in the arm. Urton hung onto the gun and ran back to the car and jumped in with the others. My dad said that you could see the marks where they spun their wheels getting out of there.

So they disappeared down the road and headed west across Queen Valley to Harmon's homestead. They weren't home, but Urton stopped and washed his wounded arm in their fifty gallon drum of drinking water. The shot broke his arm so it was pretty bloody. Then Urton headed on into Banning. When Harmon got back he couldn't imagine what had happened.

So late that night, the deputies came out to the ranch and arrested Dad and took him in. He got out on bail, and it was several months before the case came to trial.

Dad had a good lawyer, and he had good evidence. Many people had heard Urton make threats that he was going to get my dad. The trial lasted four or five days. It was proven that my dad had fired in self-defense, so he won the case and was acquitted.

Large rock cattle tank at Quail Springs provided water for cattle and homesteaders.

CATTLE DAYS

Lost Horse Valley, Queen Valley, and down around Quail Springs were mostly the summer range. Later when Sheriff Stocker and others were running cattle too near Quail Springs, we kept away from them as much as possible. In winter time we'd run the cattle over to Pleasant Valley and down in that area during the coldest months.

When we had cattle in Pleasant Valley, we'd water from the Pinyon Well. My dad had an agreement with Fred Vaile who had the Eldorado Mine to use the well at Pinyon and the old Eldorado pipeline. My dad set up a pump there, and pumped water down into the valley about two or three miles through the pipeline that he fixed to a water trough. Because the water in the trough only lasted a week, it had to be pumped every week. It took two or three hours or half a day to do it.

Dad kept his cattle there most of the winter because it was usually pretty good feed down there and it was warmer. He tried to keep that Pinyon Well in operation so our cattle would have plenty of water. He also watered our cattle at Stirrup Tanks. Dad had at most about 65 head of cattle at one time. He didn't feel that he could run anymore on the range and properly feed them.

Dad also used to go out here in the rocks at the ranch and pull bunch grass. He would take a big gunny sack - two or three of them in fact - and stuff a sack full and then get another one and stuff it full. That was the evening's feed for the horses. He would also get up here in the rocks where the horses couldn't get to graze themselves and he would pick that grass. He also used to haul hay up through the old Berdoo Canyon. He'd go out to Pleasant Valley and down through Berdoo Canyon into Indio. That could be a real experience at times.

We also ran cattle for several miles behind the dam where there is one little valley opening into another all the way to the old Wall Street Millsite. You have to climb over a few rocks to get to the millsite, but there are quite a few openings. We called this the "back pasture" and kept it mainly for horses which we turned in there. This was a good place for them to graze and we could easily go out and find them. We didn't have to spend a lot of time looking for them.

This area is naturally fenced by rock. To get the horses we might have to walk a mile or two, but that was better than walking three or more.

There also used to be good stock grazing in Queen Valley. Whether it got a little more rain there or not, I don't know, but the grass usually seemed to be a little bit better up in that area. The little dry lake over there just off the present Park road behind Elmer's Tree used to be a lot more prominent than it is now.

Most of our branding was done at the ranch in the corral there. Occasionally we'd do some out in the field, and we had a corral out there by the Barker Dam where we did some. Then over on the old Bill McHaney homestead we had another corral there.

In 1935 I went to Ontario to go to high school and was only home during the summers. At that time it was my chore to take care of the cattle. I'd spend more time herding them than doing anything else.

The day after I got back from high school, I'd be out on horseback, and after the first week, I'd be so sore I could hardly walk when I got off. After that I'd toughen up and have no more trouble.

At that time we used different springs for water and kept moving the cattle around all of the time. We'd water at Stubby Springs and Quail Springs, and then over in the Queen Valley. There would sometimes be water in that little tank that my Dad made on the summit just before you go down into Pleasant Valley, and we'd head them back in there. We'd check the water every couple of days to make sure that there was enough.

We never used Pine Spring for cattle watering regularly. We did water there once in awhile, but only when there was nothing else available in that area. It was a tougher trail for them to walk down. Sometimes we'd haul water for them.

When the tank would go dry out there on the summit junction of Queen Valley, we'd haul water over so that we wouldn't have to move them, and they could stay in that area.

In late October of 1943 my mother put some of our cattle up for sale. She sold 61 head to Henry Paul and Harold Lienau, partners in the B-bar-K Cattle Company of Morongo Valley. The sale was to help pay Dad's defense attorney after his trial.

I remember going out and helping round them up with a buddy of mine - Lee Potter from Alhambra - and Uncle Aaron. We spent all day one day getting them over to where the Park Service put their entrance to Barker Dam. That used to be our holding corral and we used it for our roundups. We'd drive all of the cattle in here, and we had a chute to load them into a truck.

Well, we were within a quarter mile of the corral when it got so dark that we lost them. So the next day we had to go back and round them up again.

This time we got them into the corral just as the trucker came up to haul them. We had some Brahman cattle in this herd, and they were hard to get in that chute. Two or three times those darn things ran us out of the corral.

I remember an earlier time when my dad butchered a big old steer near that live oak tree to the west of the present parking area. He had given this steer to my brother just about when he was born I guess and planned to butcher him when he came to be four or five years old. I think the steer was about eleven or twelve years old when my dad decided to finally butcher him. My brother had since died, but Dad said that he didn't know what it would be like being that old but "we've got to butcher him anyway." So we did. He was a great big one.

My dad had a tripod that we just put on the back of the truck. He would unload it and set it up using a chain block and a spreader bar between the steer's hind legs. Then we would lift the carcass up. And as you skinned, you raised the carcass a little higher

It was amazing to think that he weighed a thousand pounds dressed. So that was a lot of meat. And you know that meat was some of the best meat that we ever had.

That was just that one time that we butchered there by that oak tree. Normally when we did butcher beef it was at the ranch. But we really didn't eat a lot of our cattle. We would butcher once in awhile, and we'd do without beef for awhile. When I was a kid here, I'd go hunting rabbits or quail which my mother would really fix up well.

The next morning after Dad butchered, we could count on liver for breakfast. We'd have liver and onions, as well as bacon and eggs. Brains was one of my dad's favorite dishes, but he also liked sweet bread, heart, and tongue.

My mother would bake the tongue. It made very good sandwiches, but I didn't care for heart or sweet bread or brains.

When dad butchered, he saved just about everything that he could. Even the head was used. He would take that after he had skinned it and taken out the tongue, and he would season it. He used onions, salt, pepper, and garlic and wrapped it in a flour sack with the meat and everything still on it. Then he would wet it down real good, cover it with two or three layers of burlap and wet that down too.

Then we'd build a fire out in the wash and throw a bunch of rocks in and keep that fire going almost all day. That evening after it had died down, Dad would dig out all of those hot rocks, put that head down in there, and place the rocks in on top of it. He covered the whole thing with hot sand and left it until the next day. And you know that was some of the best flavored meat that I think I ever ate.

When we did butcher we kept the beef in the old tunnels over in the bank below the present south dam at the ranch. My dad would hang the beef out every night after sundown, then get up early in the morning before the flies came around, and wrap it in three or four layers of canvas or burlap and put it in the tunnel. He did that every day. He could keep meat for a long time that way. That's the way you had to keep it, because we had no refrigeration.

But in the summertime when it was pretty warm, he would keep an eye on it. And he would finally say, "Well, it's time to can the meat."

So we'd get it out here and everybody - all us kids and my mother - would cut that beef up and pack it in fruit jars and can it. It was good meat.

After the tunnels washed out in the early '30s, we had to find other places for storage. We used to put some of the stuff over in what we called the "assay cave" over there across the road.

Then my dad built the shelves in the machine shop, and we had a lot in the barn too. But he had a lot of other stuff in the barn and didn't have too much room for canned goods and stuff.

He and my mother canned two to three hundred quarts of fruit, meat, vegetables and stuff like that every summer, and we had to have a little storage for that. So he built this shed, and he had quite a little storage there. But we missed the tunnels because they kept everything at a pretty even temperature year round

Barker-Shay cowboys below Barker Reservoir in the early 1920s.

MAKING RAWHIDE

When Dad butchered beef, he always kept the hide. He would take it out probably the next morning after the butchering and dig a hole out in the wash down below where the old well was. Then he would take a couple of wheelbarrow loads of ashes that he had saved from the fireplace down there too.

Well, he would wet this pit down, dump in the ashes, put the hide in, cover it up with ashes and sand, wet it all down good, then leave it for a week or two. That would take the hair off. The ashes acted on the hair.

After a week or so, he would dig the hide out, because it would be getting pretty rank by that time. Then he would stretch it out on the ground and scrape the hair off. He'd use a scraper for that which was a piece of metal like a big putty knife. Using that the hair would slip right off.

After that he would stretch that hide out on a level stretch of ground and stake it out. He would stretch it and stake it out all the way around so it was tight.

Then he would get in the middle with his knife and start cutting. Usually he cut a stripe about 3/4 of an inch wide as he went around in a circle. Around and around and around he went until he had cut that whole hide up into that 3/4" wide strip of rawhide.

After that he would take the strip and gauge it. That is, he would set a knife in a notch of a piece of wood at the width that he wanted to cut. The knife was sort of in a Y-shaped setting. Then he would pull that strip of rawhide through it, and it would trim off a slice of it and bring the rawhide's size down to about 5/8th of an inch wide.

Then he would turn it over and do the other side of the strip the same way. Finally he would end up with a strip of rawhide about a half an inch wide.

When he was done, he would just roll it up in a big roll and hang in the barn, and it would get just as hard as a board. When he needed it, he'd take it down, throw it in a bucket of water, and let it soak for two or three days. Then it was pliable again and ready to use.

He used it to make harness parts, and stuff like that. I remember particularly when we were running the mill, we had a lot of wide belts. And you could buy metal lacing for a belt, but on those wide belts, it didn't hold up very well.

So my dad use to lace them with this rawhide. He would punch two or three rows of holes across the belt on each side where he was going to bring it together, then he would lace it. There were actually three rows of lacing when you finished up. And that would hold for a long time. Those belts, which are still up there at the mill, were a composite of rubber and canvas bonded together like the carcass of a rubber tire. There's probably some of that rawhide that he made up there too.

One thing about rawhide is that it's pretty limber when it's wet. You can wrap it around something and make a tie and that stuff will dry and shrink and tighten up as tight as a banjo string. So it was useful for a lot of things around the ranch.

Willis Keys holding a gold pan about 1950.
Photo by Paul Wilhelm.

THE COLD WALK BACK

Usually we used to go down to Pleasant Valley in the spring and cut grass by hand with sickles. My younger brother, Ellsworth, my dad and I would go down there and camp two or three days with our old Maxwell truck. The grass used to get up high when there had been quite a bit of rain that year. And we'd pile it up, bring a whole truck load back to the ranch, stack it in the barn, and use it for feed.

Probably around 1932 or '33, Dad was running the Wall Street Mill twenty-four hours a day, and whoever was having their ore milled would stay there too. One would run the mill during the day, and the other would run it during the night. So my Dad couldn't leave to get hay, but he would come to the ranch and check on everything once in a while.

My uncle Buster was staying at the ranch at that time, and he was taking care of most of the ranch chores. When we ran out of hay, he said that we'd just have to get the old truck and go to Indio and get it ourselves.

So we did. We drove the old Maxwell which had a windshield but no top, and we went down through Pleasant Valley to Berdoo Canyon. Those were all dirt and sand roads. In fact, there wasn't a real road down Berdoo Canyon. It was just two or three car tracks down there then.

So on we went to Indio, went out into the fields, picked up the hay bales, and loaded them on. I think that we loaded about a ton or maybe a little over before we headed back up the canyon.

By the time that we got up into Berdoo Canyon, it was getting dark. There were no headlights on that old truck, and all we had was a kerosene lantern. The tracks just wandered in and out through the brush, so my uncle told me that I would have to walk ahead with the lantern and pick out the road for him. I did this for about two or three miles up that canyon. Finally the canyon got narrow enough so that we could tie the lantern on the front and go ahead. We got up to that last pitch which we called "the waterfall", because it was so steep. Well, the old truck just wouldn't make it, so my uncle said that we'd just have to unload enough bales so that it would.

We threw off about two thirds of them, and the old truck made the incline. Then we had to go back and drag those bales up and load them back onto the truck. It started to snow about that time, and it sure got cold. It was a real chore to get loaded up again so we could get moving, and this was all in the dark.

By now the snow was coming down like a son-of-a-gun, as we

went across Pleasant Valley. On the truck we had a big old canvas tarp which my uncle pulled up over his feet, and I pulled it up over my head. I also had this big old overcoat of my Dad's to put on, and I soon fell asleep under there.

It wasn't long until Buster woke me up and said, "Willie, I think we're on the wrong road." Then he added, "I think we took the old Lost Horse Road."

I looked around and didn't recognize anything. So we turned around and sure enough he had taken that road where you leave Pleasant Valley. The old back road to Lost Horse went up there to the left, and he'd taken that and gotten up in there by mistake.

So finally we got back to the right road, turned, and made a couple hundred yards before we ran out of gas. And it was still snowing. We left the truck and took off walking up the road with the lantern and headed towards the mill. It was close to two o'clock in the morning, but we walked and walked. The old Mohair coat that I was wearing got soaking wet and weighed about two tons. I wanted to get rid of it and take it off, but my uncle wouldn't let me. He was afraid that I'd freeze to death, but that coat was wearing me out.

Finally the snow hitting the glass of the lantern caused it to break, so we had to put it out and walk in the dark. The only things you could see were the dark spots in the road where the snow had melted leaving puddles. I remember the dogs barked at the Harmon place as we walked by. Then at daybreak the next morning we walked into the mill. Dad had just gotten up, and I just climbed into his bunk while Buster climbed into another one. I guess I slept all day.

Dad took a couple of gallons of gasoline and walked all the way back down to the truck, got it going, and brought it back. He didn't return until sometime that afternoon which was when I woke up.

Ruins of the bunkhouse at the Wall Street Mill in 1976.

BLACKSMITHING

In the early days every mine or cattle camp out here had a blacksmith shop, even if it was just outside and up against an old oak tree or built in the rocks. Most would have a forge with a big old bellows on it so they could do their horseshoeing and fix their wagons or different parts that broke. They had tools, and if they didn't, they made them. Over the years, my dad collected a lot of these tools. He had hundreds of pounds of them around, most of which he used.

Coke was the preferred fuel for the forge because it was a little more readily available out here. It also made a better fire and didn't make much smoke. Dad also used coal and bought it by the fifty pound sack. But coal was pretty scarce, and when we ran out, we would gather what we called "petrified yucca". That's dried pieces of Joshua tree.

In the '30s the old prospectors would bring their tools to the ranch for Dad to sharpen. After a pick had worn down, you could keep sharpening it and retempering it, but finally it would get too short.

So Dad manufactured points that were split in the back. He would take the old worn pick and sharpen it down and make kind of a wedge out it. Then he would take one of his new points and just slip it on. Then he'd heat them both to the right temperature and beat them which would forge or weld them together. So my dad knew how to temp tool steel, drill steel and picks.

Drill steel had to be sharpened - blacksmith sharpened, pounded, and then tempered just right. If the metal was too soft, then it would get dull right away. If it was too hard, it would break off. So there was quite a trick to know just how to temper it.

You had to watch the color come down when you heated a piece. You'd just dip the end in water a little bit in the bit end take it out. When you took it out, you could see that heat coming down to the end of the bit. Then when it got just a certain color at the end, you would dump it in the water again to cool off the whole thing. So there was a trick knowing the color.

RAISING GOATS

I can just barely remember when my folks had this great herd of Angora goats. They raised them for the wool like you would with sheep. Just how many there were I can't remember but there were a lot of them.

Dad had a big herd and had to hire two herders who had two or three dogs to be with them all the time. They had to keep the coyotes out of them, and to keep them out of the cactus. This was probably in 1923 or '24 and probably the first time anyone had that large amount of goats out here. Johnny Kees had them in later years.

Well, the goats got fat and had a good crop of wool, but it took too much work to take care of them. The goats would get into the cholla cactus and get loaded with that stuff. And then when you sheared, which you did by hand with these hand grip shears, it was a lot of work.

But getting the cholla out of their hair was really the tough part. The wool had to be carded with a card which is like a wire brush with little short teeth. It's like a curry comb. And you card that wool, because to sell it, it has to be clean and it has to be straight.

So you card it, and card it, and lay a little bunch aside, and then take another strip. That took a lot of time. So my parents decided to get rid of them. They were herded back down to the railroad at Garnet (today's North Palm Springs) and shipped out of there.

We had domestic goats off and on for many years. These we used for milk and also butchered. It was good meat. We would only have up to a dozen or so at different times.

My folks let them run loose, and the goats would practically feed themselves. And out in the rocks where the horses and cattle couldn't get, they'd get up there and find good feed.

But we had to be careful of coyotes for them too, because they were looking for them. Of course, they would go for the young ones first if they could get them.

One time I remember my dad had been up early and had let the goats out of the corral. And they had gone down where those archaeologists were digging this summer in that flat place beyond the old mill. Well the goats were all in that area feeding a little bit. My dad had come back to the house when he heard some commotion down there with the goats. So he grabbed his rifle and ran right back, and there was a coyote in the middle of them. I think he had killed five goats before my dad shot him. He had bitten them in the neck, dropped them, then gone on to the next one.

That was one of the hazards of raising goats. They could protect themselves pretty well if they could get into the rocks. They could get up onto places where a coyote could not actually get. If a coyote caught them down on the level ground, they were fair prey. So we tried to get them in every night and put them in the corral. And if they were out late, we'd have to go out hunting for them. Usually two or three of them would have bells on, and if we had any luck at all, we could hear the bells when we got close.

Left to right: Archaeologist Cindy Stoddard screens material during a University of Nevada (Las Vegas) dig at Keys Ranch during the summer of 1996, while Nelda King, Corinne Keys, Willis Keys, and Harmon King look on.

FAMILY FRIEND - BILL McHANEY

Bill McHaney built the adobe barn in 1894. He came out here in the 1880s with cattle and was one of the first cattle men out here. He told us that when they came over the pass in today's Yucca Valley and down into the valley that it looked like a wheat field. The grass was three feet high. It was really amazing, and there were herds of antelope all down through here.

Originally there were two adobe cabins down below the barn in the flat area. Dad took them down, saved the adobe, and started to build an adobe house over there across the wash. He put in a fireplace but no roof. Then he decided to do something else, and let it go. And of course, it all melted down. The adobe didn't stand up to the weather without a roof.

The original strike for the Desert Queen Mine was found on the top of the canyon by a little pinyon or a juniper. Bill said that it was found in a rat's nest under a tree and that's where the ledge of really rich ore showed up. And then they worked their way back on that ore, taking it out, and they ran into this flat spot. It was a smooth curved rock. And Ryan told my dad, "I went over there when I heard about their strike and looked. And that smooth rock was plated with gold. It was just covered with gold."

James found the original rat's nest discovery. He had brought some of the rock to Lost Horse and some of the McHaney boys were there and saw that. Then they followed him back to his strike.

I heard Bill tell a story about the shooting of James. He said that James was killed in self-defense by the cowboy that shot him, but then he was one of those involved in the shooting. Bill said it was a fair fight. Their version of what happened was accepted by the law and nothing else was done about it.

There was a kid who was cooking for the McHaneys by the name of Delong who said, "Those guys play pretty rough. One night they were camped and had finished supper when one guy jerked out his pistol, fired it, and ran behind a rock. Then he shot a few rounds over these guys' heads and the rest of them scattered in the rocks and were shooting at each other." Well, that's how rough these guys played.

The Desert Queen Mine was strictly a pocket mine. After the McHaneys hit a pocket, the vein would pinch down to 2" to 1" to nothing. Anytime they found a little ledge or something they dug into it. If it started looking good, they went further.

When the original mill was built there at the ranch, it was financed for the McHaneys by an outfit from Los Angeles to get them started. The Baker Iron Works built the mill and ran the mine until

56

Bill McHaney and his long-time friend, Bill Keys, about 1932.

the mill was paid for. They manufactured the mill, came out and set it up, and put it into operation. After it was paid off, they turned it over to the McHaneys, and things sort of went to pot. They got all of this money and didn't know what to do with it. They spent it just as fast as they got it.

After they left, it went through several hands. Morgan was the last one. Morgan hired my dad to do the assessment work on the mine and to live there at the ranch and to take care of it. This was suppose to be done for a certain amount of pay. Well, Morgan never paid him, and my dad kept after him.

Finally Morgan died broke and left no money to pay Dad. So his business manager said, "Why don't you just file on the 5-acre ranch for your pay," which my dad did. He also located the mine, and that's how he acquired both.

A few years later Dad filed for a larger homestead which included the ranch and the Cow Camp area out to where the fence line is now. That was the original 160-acre homestead.

If you had proved up on your homestead, you were allowed to

take additional property through the Homestead Act. So later he took up a couple of 80 acre stock grazing deals that the government had at the time. And then later, he of course, bought up some homestead relinquishments, and later he got my grandmother's homestead.

After the McHaneys lost the ranch and all the money was gone, Bill ended up with nothing really. So he went back to prospecting. Years before when he and Jim first came out here, some old Indian had told him about this place over in Music Valley. There was a rich mine on that hill.

Bill was really convinced that this old Indian was telling him the truth, He trenched every fifteen or twenty feet all around that mountain by hand looking for that outcropping. He spent most of the rest of his life which was about forty plus years making this his project. He never found anything, and to this day I imagine that you can still see the trenches that he dug.

Of course, he didn't work at it steadily, and in later years when he got more feeble, he spent quite a lot of time over here at the ranch. My parents took care of him.

In the summers when the weather was half way decent, he'd work there in Music Valley and then come over to the homestead or to the ranch in the wintertime. He'd dig all day and come in about sundown or dark. He was pretty active considering that he couldn't see very well. He had sun stroke many years before, and that had left him about half blind. He also had rheumatism pretty bad, and when the weather got too hot, my dad would go and get him and bring him back over here to the ranch.

He was in his seventies then and pretty feeble, and he had this little shelter there near his diggings. It was only about four feet high, and you had to duck to get into it. It was just big enough to have his belongings in there and a little bit of grub, but he had everything arranged. Bill also had a little rock fireplace with a grill on it, and he had a little seat there.

On one trip when my dad drove over there with Oran Booth, Bill told them about something that had recently happened. He said that one evening he had gotten his fire going to cook some potatoes. He knew he had a couple of them left, so went into his shelter, reached into this box and felt something.

Bill told them, "I knew it wasn't a potato or anything because it moved a little bit." So he took it and threw it out the door. It was a big rattlesnake.

There were several fellows working in the Gold Park area at that time. One by the name of Jenson had claims there for several years. He would take Bill to 29 and see that he got groceries and stuff.

Bill also had a cabin over near the present day parking lot at Barker Dam. My dad had bought the relinquishment on this property from a fellow by the name of Smith who had started a homestead on it but hadn't proved up. There were still a couple of buildings on it. So to make sure he had title, he had Bill stay there and had him homestead it. He stayed there five or six years. At his death Bill willed it to my dad.

As Bill got older and more feeble, he used to come to the ranch and stay. My dad would go over to his cabin or to Music Valley and pick him up and bring him over for the winter because the winters were getting pretty severe. He liked to come over to the ranch and get some regular meals. We had a little cabin fixed up for him where eventually he died.

Dad also rigged an intercom system between the house and the cabin so we could talk with him and find out if he was okay. We could call him and tell him that we were bringing over some breakfast or whatever. And I used to go over there and take him his breakfast or whatever they sent me over there with, and he would sit down and tell stories. I was probably eight or nine years old, and my sister Virginia was about seven or eight.

We would sit out in the orchard and sometimes the sun was shining and we were sitting under one of the trees. He would tell us these stories about the Desert Queen Mine, old mining people, cowboys, and the old times.

I heard these stories and to a kid they didn't mean too much. Now I think back and wish that I could have really listened to those

Bill McHaney in front of his crude shelter in Music Valley about 1933.

stories so that I could have remembered them now. Bill had lots of stories. He could sit down all evening and tell about the old days. It's a shame that those have been lost.

I could have sat down with my dad too and asked him questions in the later years, but I didn't do it. I always thought I would have plenty of time. So much of those are lost too. I could have asked him about the stories that Bill McHaney told. Some of the other old-timers out here sat in the front room at the ranch and told stories to him.

But I can remember a couple of lines of a song that Bill McHaney used to sing. It was something about,"Over Jake we held a wake in the days of '49, and he ran into a knife in the hands of old Bob Kline." As far as I know that was a song about some of his own experiences. These people--Bob Kline and Jake--were all real people.

While Bill was at the ranch, he always tried to do something. My dad was raising alfalfa and a garden, and Bill would be out there shoveling and watering and cleaning out the watering ditches and stuff like that to keep busy all day long.

But he couldn't hardly see. He was almost blind. He could just see an outline and almost to the last he kept his old shotgun. It was an old 16-gauge, long-barrel, single-shot. He could get a rabbit once in awhile when he was over there at his camp. But he got so that he couldn't' hit a rabbit even with his shotgun.

I used to hunt rabbits then quite often, and I would go out each evening or so and get a couple of cottontails. Bill liked that. He really enjoyed cottontails.

So this one time I was getting ready to go down and hunt for some rabbits and I got my .22. Bill was out there at the ranch outside his cabin when he saw me leaving.

He said, "I'm going to give you this shotgun. I can't see well enough to shoot anymore."

He had given Dad his rifle years before, but he had kept the shotgun because he had enough sight to get a rabbit once in awhile until now. Well, he handed me his shotgun and a whole flour sack full of ammunition. I was about ten or eleven years old and so proud of that.

Then he said, "Now when you shoot this thing, hold it tight against your shoulder. It kicks pretty good."

So I headed down the road and a big jackrabbit jumped up. Without thinking, I wiped up that shotgun like I would a .22 and pulled the trigger. I think the hammer or something came up and hit me in the nose and bloodied it. So I always remembered after that to hold that thing tight against my shoulder.

In the days that I knew him, Bill was a very gentle fellow. He never spoke a cross or a bad word about anybody no matter who it was. That probably was a change from his earlier life, but maybe not too much. He may not have been as wild as his brother or the rest of the bunch.

Physically he was rather stooped and rather slim. He had a little bit of hair but was pretty well bald. He was also very wrinkled and usually had a couple of days' growth of beard. He did shave occasionally. Bill mostly wore an old felt hat and always wore a bandanna around his neck - either blue or red. He always had that. He wouldn't come out of his cabin unless he had that on.

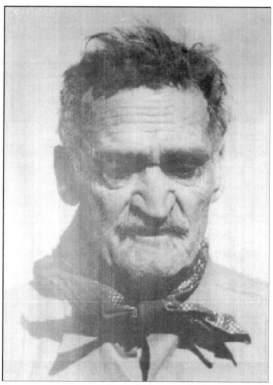

Bill McHaney in 1935 wearing bandana that Willis Keys remembers.

BILL McHANEY

Found and Lost a Gold Mine

Driving into 29 Palms from the east, motorists are greeted by the community's seventh mural commemorating its early history. Artist John Whytock's Dirty Sock Camp depicts those early mining days when countless individuals roamed the desert and surrounding mountains in pursuit of gold.

One of the most famous of those miners is fittingly depicted in this mural. History remembers him for being one of the earliest residents of the area, for having had one of the richest mines, and for being late to his own funeral. Early cattleman, miner, and much later, beloved member of the 29 Palms community, Bill McHaney was a legend in his own time.

Born in Gallatin, Missouri on March 25, 1859 as William Lafayette McHaney, he came across country by ox-drawn wagon with his mother, brother and two sisters when he was eighteen. First stopping in Tulare, they soon headed south settling in Santa Ana Canyon area below Big Bear where the brothers started a cattle business.

Within two years they moved their operation to Thousand Palms and increased its size by bringing cattle from Arizona. More range was available in the higher mountain valleys to the north, so they relocated there settling in a rock enclosed canyon where water was available. This site would later become Bill Keys' Desert Queen Ranch in today's Joshua Tree National Park.

Bill McHaney said that when he first came to the area in 1879 "the valley was full of antelope" and the area's lush grass was "belly-high on a horse".

The Langs, cattlemen from the Indio area, soon became their neighbors. In 1893 while looking for a lost horse several miles from the McHaneys ranch, John Lang discovered a chuck of rock that contained gold. So he and his father supposedly gave the McHaneys 75 head of cattle for the south half of their range containing 2 waterholes and filed on the claim.

The resulting Lost Horse Mine became prosperous, and the lore of gold caused the McHaneys' interest in the cattle business to wane. They soon began looking for their own strike - one that would equal the Lost Horse's success. The following year, Bill McHaney would be lucky.

In recounting its discovery, Bill said in 1935, "You know, I discovered one of the biggest gold mines in these hills. In the summer of '94 I was over in Queen Valley about 12 miles south of here and run into some rich float. It took me five months to locate the real deposit. On January 24, 1895 I found her and I called it the Desert Queen Mine."

They first hauled their ore for processing to Pushwalla Canyon and the Pinyon Mill, owned by Tingman, Holland, and Tallant. From their profits they went to Los Angeles and bought a five-stamp mill to do their own processing at their ranch. The mill was setup on the banks of the wash, and the two brothers made their sister, Carrie Harrington, a third partner.

A July 20th, 1895 article in a Riverside paper described their new venture: "Mining is booming in the Pinyon district. McHaney's new five-stamp mill has just commenced dropping on the rock of the famous Desert Queen. After running three hours quite a thick coating of amalgam was perceptible on the plates. On the average the ore will mill $200 per ton. The mill is a beauty and is complete in every particular. The Baker Iron Works of Los Angeles were the builders, and their superintendent, Walter Small, placed it together. It will break 15 tons of rock every 24 hours. There are twelve men working on the Desert Queen, and about the same number hauling ore and wood and working around the mill. Altogether, McHaneyville is quite a camp."

Unfortunately poor management and wild spending by Jim McHaney soon caused the brothers problems. Lingering debts owed for supplies and materials as well as for the wages of their workers were not paid. Sister Carrie Harrington secured a bank loan and came out to the desert to oversee the operation. When profits continued to fall, the Zambros Bank of Los Angeles took over the mine and the mill site, and the McHaney brothers were out of the mining business.

This 1914 photograph of the Desert Queen Ranch taken by Bill Keys shows the adobe cabin and barn built by the McHaney brothers almost 20 years earlier.

Top: 5-stamp mill built in 1895 by Baker Iron Works when the McHaney brothers operated the ranch as a mill site. Photo taken about 1941 before parts were sold for scrap metal.

Left: 1930s photo of Bill McHaney with his shovel and the shotgun which he later gave to Willis Keys.

The bank appointed a succession of receivers to take over the property and manage the mill, and while one was able to show a small profit by reprocessing the dump tailings, the mine seemed to run out of ore.

Jim McHaney, who was supposedly driven from the ranch at gunpoint by his sister, moved down to the Coachella Valley, got involved in a counterfeiting scheme. On September 8th, 1900 he was found guilty in Los Angeles US District Court and was fined and sentenced to 3 years in San Quentin. After his release, he settled in the San Fernando Valley and eventually went to work for the Los Angeles Water Department.

In 1932 he stopped by his old ranch and mill site, now Bill Keys' Desert Queen Ranch, on the way to see his brother, Bill in Music Valley. Willis met him that day and told me that Jim died sometime within the next two years.

Bill McHaney left the ranch to continue prospecting in hopes of finding another strike as rich as the Desert Queen. From his friend, Captain Jim Pine, leader of the Serrano Indians living at the 29 Palms Oasis, he learned that there was gold in the hills to the south and east.

About a week before Jim's death, he told Bill how to get there. "Captain Jim's directions were as clear as could be. I made a camp in Music Valley in 1905 - over 30 years ago - and I am sure that is the valley Captain Jim referred to. I've been going between that camp and my home place here ever since, packing in grub and water and bringing out a little ore now and then. I'm just about to strike pay dirt now," McHaney related.

He spent those years living part-time at the Oasis and in a small dugout shack in Music Valley. Early miner Oran Booth remembered visiting that camp in the fall of 1932 when he first met Bill McHaney. "He had the poorest camp that I ever saw anyone have. He had a double bedspring which stood up in some rocks about three feet off the ground. Above that he had some sticks stuck up and something over the top of his bed to hold the five-gallon tin cans and paper cartons that he used as a roof. Out in front he had three little rocks over which he built a fire to cook on. He had two maybe three stew kettles and a couple of frying pans setting on the ground."

Until his later years Bill McHaney continued to spend most of his time in Music Valley trenching up and down its hillsides in his elusive search for gold. Bill Keys often drove out there to check on his friend, and whenever he was ill, Keys brought him back to his ranch to recuperate. McHaney often spent the entire winter there and often visited at other times.

In order to be closer to the ranch, Bill Keys eventually encouraged McHaney to file on a homestead near Barker Dam, and to live in the small cabin there during the winter. On July 12, 1933 he legally filed on the one hundred sixty acres and stayed there for about two years before moving permanently to Keys Ranch due to poor health.

Four months before his death, he came to 29 Palms with the Keys to see Frank Bagley for the purpose of making out his will. He told Frank, "I am leaving everything to Bill and Mrs. Keys. They are the very best friends I ever had, and have taken care of me as no one else would."

Bill McHaney died of pneumonia on January 5, 1937 in his cabin at the Keys' Desert Queen Ranch. His funeral was scheduled to be held two days later at 29 Palms' new cemetery, but deep snow prevented his trip from the ranch. When the home-made coffin bearing his remains was brought down the next afternoon, the town's people turned out to say farewell.

A few weeks later, Bill Keys carved a stone monument to mark his friend's resting place. Today sixty years later, some of his trenches in Music Valley still remain as silent testimony to Bill McHaney who found and lost a gold mine.

TWO SUITCASES OF GOLD ORE

Defaus Geil came out in 1923 or '24 and homesteaded in Morongo Valley. He also came up and leased the Desert Queen Mine from Dad. He didn't know anything about mining, but he wanted to get into it anyway.

He and his wife went up there and worked for about a month using a hand drill and a hammer. Then Geil told his wife, "This drill won't go in any further, and I don't know what's wrong." He didn't see too well, because he had only one good eye. So he put in a little charge of powder and blew it open to see what it looked like, as the little hole was only about eight inches deep.

Well, after the explosion, the rock sort of bent out. It was all tied together with gold. Here was a little pocket of gold that had been left over from one of the old original pockets that had been missed by earlier owners.

So they dug it out, and there was enough to fill two suitcases. Later that night they came by the ranch and told my dad, "We're giving up the lease. You can have it back. We've got all that we want!"

There was $25,000 worth of gold in those suitcases. And with the money that the Geils got for that gold, they went back to Morongo Valley, built the Morongo Inn, and set up business there. It was a big rock house just at the top of the hill before you head down the grade into the Devil's Garden.

Although he gave up mining, Geil was still interested in it. So I don't know if my Dad gave the old Tully Mine to him or sold it to him. He took it over and used to come up every year to do the assessment work. But he never really took anything out of it.

Helen and DeFaus Geil brought rock from throughout the desert to build the Morongo Inn, financed by their rich strike at the Desert Queen mine. They were helped by Ray Bolster who built Bill Keys' chimney. Frasher photo courtesy of Lucille Frasher.

WHO REALLY DISCOVERED THE DESERT QUEEN CLAIM?

In an interview in 1935 Bill McHaney took credit for the discovery of the Desert Queen Mine. "You know, I discovered one of the biggest gold mines in these hills. In the summer of '94 I was over in Queen Valley about 12 miles south of here and run into some rich float. It took me five months to locate the real deposit. On January 24, 1895 I found her and I called it the Desert Queen Mine."

Despite Bill McHaney's statement, the real discoverer of the claim was a worker at the Lost Horse Mine by the name of Frank L. James. The 35 year old miner was six feet tall, weighed 175 pounds and had dark blue eyes and brown hair. Unfortunately for him, he did not profit from his discovery. Instead he died at the hands of Charlie Martin, a cowboy working at the McHaney ranch which later became Bill Keys' Desert Queen Ranch.

Riverside and San Bernardino newspapers all carried accounts of the killing which occurred on April 5, 1894, "on the desert 40 miles north of Indio." Immediately thereafter Charlie Martin appeared before authorities in San Bernardino to inform them that he had killed James and that he had acted in self-defense.

His account in the *RIVERSIDE DAILY ENTERPRISE* stated that he had left San Bernardino six weeks earlier with a grubstake to prospect, and on the 2nd of April he had located a quartz claim which appeared to be quite rich. He said that two days later James supposedly located the same claim.

When Martin met James on the site of the claim the following day in the presence of George Meyers and James McHaney, Frank James asked Martin if the claim marker on the site was his.

Martin said, "Yes," and supposed James told him, "You go and tear it down or I will cut your heart out!"

Martin replied: "I will not tear the monument down and you will not cut my heart out!"

Then James supposedly made a lunge at Martin with a knife cutting a 3 inch gash on his left arm near the shoulder, a 3-inch gash across his chest, and slashed his left hand "which he was using to ward off the blows of the knife."

"A gun belonging to one of the other men was lying on the ground several feet back of where the men were struggling. As James kept advancing with the knife, Martin grabbed the gun and shot while he was advancing. This did not stop him and another shot was fired, taking effect in James' breast killing him.

"The miners in the locality then congregated, examined into all the circumstances and the meeting concluded to bury the body owing to the heat of the desert and the long distance to the railroad. Immediately after the tragedy Martin went to San Bernardino and the coroner's and the sheriff's office was notified. The men who witnessed the tragedy are there ready to testify.... Martin was photographed without clothing so as to preserve the evidence afforded by his wounds of the truth of his story."

San Bernardino authorities sent a county surveyor to the killing site to determine if it took place in Riverside or San Bernardino County, and a coroner also went to exhume and examine James' body. He also examined Martin's wounds before making the trip.

When the coroner returned the following Tuesday and an inquest was held on April 16th in San Bernardino. The verdict of the jury was that Frank L. James "came to his death from two pistol shot wounds inflected by Charles Martin. And we further find that Charles Martin acted in self defense while his life was in jeopardy at the hands of the deceased."

No mention of the killing was made a year later when a Riverside newspaper article mentioned that a new strike "located by the McHaney brothers was averaging $250 a ton. James B. McHaney was in Los Angeles ordering a 5-stamp mill for $2,000 gotten from 7 tons of ore." Wagons brought the ore from the Desert Queen Mine to the McHaney ranch where the stamp mill was set up along the banks of the wash.

Almost seventy years later, when Bill Keys was answering questions about the area's early history, he recounted a different version of how the McHaney brothers acquired the original strike for the Desert Queen.

Bill said, "James was a miner working in the Lost Horse Mine, but on Sundays he'd walk and prospect. Well, these coyote cowboys (George Meyers, Charlie Martin, and James McHaney) saw his track because it was easy to follow. They were sharp as a coyote as to any disturbance in the ground.

"So these three fellows came here on horseback and stayed up on a hill above James' cabin off the Lost Horse Mine road. Charlie Martin walked down a little ways and yelled to James who was in his little cabin in the ravine."

"Do you own this? We found a rich strike here?"

"And he (James) went up to look, and Martin shot him. That was to get the Queen Mine. That was up on the ridge above James' cabin that they shot him."

Bill Keys further said that McHaney gave George Meyers a herd of cattle for his interest in the shooting and Martin was paid $47,000 out of the first mill run of the Desert Queen for his interest.

While the exact events that led to Frank L. James' death at the hands of Charlie Martin will never be known, his strike which the McHaney brothers named the Desert Queen Mine, became a producer of gold ore for them and subsequent owners, including its last owner, Bill Keys.

A SHOOTING ACCIDENT

Before I was born there was an outfit that had a mine near the El Dorado, and they came out one year in the spring to do the assessment work. The four men were unhitching their mules, unpacking the wagon, and setting up camp.

One of the men had a Winchester .44 rifle that used shot cartridge. That was a hollow wooden bullet filled full with bird shot for snake purposes. Well, he grabbed the muzzle of his rifle to pull it out of the wagon, and the hammer caught on something, pulled back, and fired. The bullet traveled through his arm and lodged near his shoulder.

Well, I don't suppose they had much in the way of first aid. They washed it as best they could and stayed overnight. The next morning the man's arm was swollen up and in bad shape, so they loaded up their wagon and headed back to find a doctor.

On the way they came by the ranch and stopped to see if my dad could do anything. Well, he couldn't 'cause the fellow was too far along. So they headed on to Banning, and by the time they got there, the guy had developed gangrene and died.

That was one of the hazards of the early days out here. If you got hurt, you had better be able to take care of yourself. Even in later years when we lived up there at the ranch, it was about the same thing. There were many miles to go to get medical help. It wasn't like jumping into a car now and going down there in an hour. It was a day or so trip!

THE WALL STREET MILL

George Meyers ran a few head of cattle up here in the 1880s and '90s. He must have been pretty ambitious because he dug several water wells and built several dams. The Wall Street Mill site was one of his wells originally.

Around 1928 Oran Booth and Earl McGinnis came to the area. They had a mining claim on the southeast side of Queen Valley west of Jumbo Rocks. They dug a shaft following a small vein of gold ore, but they needed water so they located a mill site at the old George Meyers well. They called their mine and millsite the Wall Street. I believe that McGinnis named it, as he was a kind of dreamer. He also fancied himself as an artist. I had one of his paintings for many years. It was done on canvas about 2 feet by 4 feet of an Indian camp beside a stream with a tepee and small campfire, and a canoe.

Booth and McGinnis built a cabin on the millsite. It wasn't a great building. It was made mostly of used lumber with tar paper on the outside and on the roof. It had a wood floor. That was something that a lot of cabins didn't have in those early days. They also built a rock fireplace for the cabin.

They redug one of Meyers' old wells and developed a fair amount of water. However, their mine didn't pan out. There was a very small vein of ore and not much gold. So Booth and McGinnis dissolved their partnership. McGinnis left the desert and Booth stayed for awhile and worked for my dad.

Then Booth had an offer of a job from a timber outfit in northern California. So he took it and stayed with them for a couple of years. When Oran left he told my dad to take over the Wall Street, so my dad did that and filed his own mill site on it. He held it until selling it to the Monument.

In 1929 a family with the name Oberer came to my dad looking for mining property. The father, Fred Sr., his son, Fred Jr., and his wife and two children. There was also a nephew with them whose name was Mattison, but they called him Matty. My dad dealt with Fred Jr., who signed a contract to buy the Gold Tiger Mine, the Wall Street Millsite, and the two-stamp mill located at that time at Pinyon Well.

The Oberers moved the mill from Pinyon to the Wall Street Millsite and set it up. This was no small chore for a man at least 65 years old, his son, and the nephew, Matty, who was sickly and not able to do much. So the two men moved the mill and set it up.

They did some mining at the Gold Tiger, made several small mill runs with that ore, and decided that they couldn't make it. They weren't able to make payments on the mine or the millsite, so Fred Jr., his wife and kids left the desert, while Matty moved to Yucca Valley where he started a little store. He did okay and was well liked by everyone.

Fred Oberer, Sr. stayed at the mill, and my dad told him that he would have to move. He said he wasn't going.

Dad didn't push it, but in a month or so he learned that Oberer had gone to Los Angeles for some reason and had left a young fellow at the place to guard it. When Dad heard about this, he got my uncle, Buster, my mother, and all of us kids, and we climbed into the old Maxwell truck and went over to the millsite.

Before we got there, my dad had told my uncle, "I'm going to take this guy out around the mill and show him the pipes and other things that have to be drained before they freeze in wintertime. When I get him out there by the mill, all of you just move his personal belongings outside and stack them on the truck."

Well when we got there, the young fellow met us in front of the cabin and said that he was taking care of the place while Oberer was away. A few minutes later my dad took him over to the mill and was gone for about ten or fifteen minutes. When they got back, everything was on the truck. My dad told him, "You might as well climb on. We're going to take you to your new home!"

So we moved him about a mile and a half east along the base of Queen Mountain to the Louise Tunnel. Here we unloaded his belongings at the entrance to the tunnel. There was quite a bit of gear, some belongings of Oberer, pots and pans, groceries, bedding, and several boxes of cigars.

The next day the fellow walked out and we never saw him again. Oberer never returned either, and the stuff that we hauled up there stayed for several years until it gradually disappeared.

Dad took possession of the millsite again and made several mill runs of his own ore. He also milled for Mitchell and McCauley. They had leased the mine later known as the Elton. But Dad was disappointed with the mill the way it was set up, because it was too hard for one person to run it. He decided to rebuild it.

A fellow by the name of Hopper, who was living at Cottonwood Springs in what is now the National Park, had a mine close by. He had taken out about forty tons or ore and needed to mill it. So he came to my dad and said that he was a millwright and could rebuild the mill the way that it should be. He and my dad made an agreement that Hopper would rebuild the mill in exchange for milling his ore.

71

Well, Mr. Hopper, his wife, and son moved to the millsite and set up camp. They lived in a tent and did outside cooking, as my uncle, Albert, and George Legates were living in the cabin at that time. Dad had hired Al, George, and Mr. Crowley to help the Hoppers as a six person work force.

I was present most of the time to run errands and to pack tools and nails. It was very interesting to me. Mr. Hopper was a good millwright and supervisor. Mr. Crowley was an excellent carpenter.

While some of the men were digging the foundation, Dad organized a safari to the Hidden Gold Mine to get timbers to build the mill. We had three trucks - an old 1918 Maxwell, the iron-wheel truck, and one we called the Jordan which was made from parts of several trucks and cars that Oran Booth had built for my dad.

As I recall, it took about two months or a little longer to finish the job. Mr. Hopper milled his ore even though it wasn't too good at $20 a ton. But he was satisfied. This was in 1932.

Dad took over the mill then and did a lot of custom milling for many different mines. This was during the Depression, and there were no jobs in Los Angeles. Some of these guys were fairly ambitious. They would come out here and either lease one of these old mines or just locate one that had been let go. Then they would get in there and dig out some ore. It was hard work, but they did it.

A fellow died at the Wall Street Mill who was part owner of the Elton Mine. My dad was milling ore for him and his partner. My dad had a little platform area there at the Wall Street that was covered or partially covered. It made a nice shade when it was hot. Well, this fellow who was pretty heavy set was sitting there in the shade when he just keeled over backwards and died of a heart attack.

Dad sent for the coroner, and he came out with a hearse to pick up the body. To keep the stretcher in place in the back of the hearse, they used these nickel or chrome-plated pegs which looked like little pyramids or wedges. So after they put the man's body into the hearse, they forgot one of these pegs and left it lying on the platform. Later my dad made a little mount for it on one of the braces, and it was there at the mill for years.

I remember two fellows who got together up there at the Elton Mine - McCauley and Mitchell - and they worked like beavers. They had an old Model T truck, and they'd take out five tons...ten tons... fifteen tons and haul it down to the mill where my dad would mill it for them at $5 a ton. They were getting $35... $40....$50 ore out of that place. So they would have a little stake and go back to LA, and spend a week or two with their families, buy some dynamite,

Wall Street Mill on right, bunkhouse on left, about 1964. Photo by and courtesy of Ross Carmichael.

some groceries and beans, and come out here and spend another month and a half or so, then take out fifteen tons. Those two fellows were there for four or five years.

My dad operated the Wall Street all during the '30s. I think that two tons a day was almost the maximum output. He always charged five dollars a ton to process their ore. He usually gave the miners their gold in the form that they called a "matte". When the gold is recovered it is amalgamated - mixed with mercury.

This process was the most practical for a stamp mill, as it is an ideal place to amalgamate. That is compact the mercury with the gold in a battery of a stamp mill churning this mixture all of the time. Some of the amalgam would settle down around the stamp at the bottom while the rest splashed out through the screen onto the silver plated plates. Anywhere from 20 to 40 mesh screen was used depending on how fine the gold was in the ore.

So Dad would put a little drop of mercury or two in once every hour or so in the battery to help capture the gold on the plate. You had to keep an eye on the amalgam as it formed out onto the plate. It formed in little ripples. If that amalgam started getting pretty hard, then it was time to put in a little more mercury in the battery. If it got too hard, then some of it would break off and trickle down onto the plates. Then it may not have enough mercury in it to catch any gold that's still running free. So you'd want to keep it so it was a little bit pliable and able to catch any other gold that might be there.

There's quite a science to it. I learned a little bit of it from my dad, and he was pretty good at it. He watched everything, especially the amount of water. You don't want too much and you don't want too little. It was kind of a slurry, and it went down the plates in a little wave. You could see each wave as each stamp pounded and splashed out a little bit. If you get too much water on the plate, it washes too much away. And of course, if there was too little, you wouldn't be getting any of it out. So you had to keep that water sometimes adjusted every few minutes. And the same with the mercury. If it started getting a little hard, every once in awhile Dad would sprinkle a little on the plate itself.

Dad would clean the mercury and gold amalgam off the plates, then squeeze it through a chamois skin to remove some of the mercury right then. The gold won't go through the chamois, but the mercury will. But not all of the mercury is removed, so you take the amalgam, wrap it in a little piece of cloth and put it in a retort. That's a metal pot with a tube coming off that you bring down into a pan of water. When you heat the amalgam in the retort, the mercury vaporizes, goes off in the tube and comes down in the water and turns to a liquid again. You keep heating that retort until no more mercury comes out the tube into the water. What's left inside the retort is called a matte. It's a kind of a porous gold.

That's the way they used to send it into the Mint - in the matte. Sometimes somebody wanted it melted down into bullion. Then he'd get it into the old forge and a crucible. And it takes a lot of heat to melt that. You have some flux in there, and you melt it. Keep it melted, and melt it quite a while, and watch it. You can see the other impurities in the gold work off. Then you pour it into a little mold - whatever size it happened to be.

You had to have a license to melt gold, and any gold that was melted down was required to be sold to the government. Usually someone would take it to San Bernardino and mail it in a package to the Mint in San Francisco. My dad would weigh it first. He had a regular set of assayer's scales that weighed in ounces or troy weight.

When Dad was milling for somebody else, those people would take care of sending it in. Because my dad had the license, they would send it in under his license and get the check back from the Mint. Then they would pay my dad whatever was necessary for milling. If my dad sent it in, then of course he would get the check. I'd say that the most bullion in those years was from five to twenty ounces or at least in that range. They weren't like the big bullion deposits that had come out of here in the late 1890s.

The Wall Street was about as efficient as a stamp mill could be. There wasn't too much that got away. If it ran 24 hours, he could process close to five tons of ore depending upon the rock. Sometimes it was less. Sometimes a little bit more. But I'd say that five tons was the maximum for 24 hours. So that was if he ran ten to twelve hours he could do two and a half tons or maybe just two tons through it. It wasn't a big money-making operation, but in those days that was all right. Out of his profits my dad paid for the gasoline, oil, and grease which was nominal in those times. And other than that there wasn't too much upkeep.

Remains of the Wall Street Mill battery that crushed ore and the amalgam table.

THE ELTON MINE

In the mid-1920s a fellow located the Elton Mine and brought a helper out with him. They didn't work it regularly, but they did the accessment work and took out a little bit of ore.

One day they were working together. The helper was down the shaft loading the bucket, and the other man was up on the top windlassing it up. Well, he hollered for this fellow at the top to haul up the bucket but nothing happened.

Luckily they had a ladder down in the shaft, so the helper climbed up and found the guy lying there. He had had a heart attack, but he wasn't dead.

His helper was smart enough to know about nitroglycerin, so he took a stick of dynamite and rubbed some of it on the guy's wrist. This kept him alive for a couple of hours, but not long enough for him to get him out of there. So he died.

I can remember his partner coming to our ranch for help, but he knew that there wasn't anything that could be done. Someone staying at our ranch went with him to Banning to get the coroner.

So after that a fellow came up there and took over the mine. I don't know if he leased the claim or the heirs just let it go. His name was Connell and he moved in there and built a little cabin. He was there for about five or six years. He really didn't do anything. He was kind of a promoter of sort, and he had several different crews come in and try to do a little placering on the stuff.

Connell had a wife who never said anything. He did all of the talking. All I ever remember was her being in the back end of that little cabin at the stove. He never took her any place. Whenever he went somewhere, he was always by himself.

Then Shorty Mitchell and a fellow by the name of McCauley came in, and I think they leased the hardrock part from Connell. They went down the shaft where the guy died on the windlass and found some pretty good ore. So they went ahead and produced there for two or three years. I'd say from about 1933 to '36 - somewhere in that era.

My dad milled the ore for them over at the Wall Street, and they got pretty good results. They got $35 to $40 a ton, and it was enough for them. After they milled 15 or sometime 20 tons, they'd go to Los Angeles and spend a week or two. Then they'd load up, buy some dynamite, beans and groceries, and come back to spend another month or two picking out a few more tons.

DEALING WITH A CLAIM JUMPER

There were times when a miner had done his necessary assessment work and someone else would locate the claim. That is they would jump right on it and locate it when the first miner still had rights to it. This was called "claim jumping".

I remember one instance when my dad ran into something like that on his Hidden Gold Mine down below Key's View. My dad had kept up his assessment work and had a buyer who was interested in it. Then someone else came over, jumped the claim, and put his location notice on it.

Well, my dad got a hold of him and somehow got him to come to the ranch. His prospective buyer was also there at the same time as well as my uncle.

Dad's prospective buyer had a big, old Lincoln. So they talked the claim jumper into taking a ride with the three of them. My uncle came along and brought a roll of rope. Out in the middle of Lost Horse Valley, they pulled up under a big old Joshua tree, stopped, and all got out of the car.

Then my dad told this fellow, "We want you to take your notices off that mining claim. You've got just two choices - Take the notice off, or we'll take this rope out and hang it over that Joshua tree and string you up here."

As my uncle threw the rope over the top of the tree, the guy said, "All right! I'll sign it off."

My dad had the papers ready, pulled them out, and the claim jumper signed off. He got out of there, and we never heard from him again.

February 1922 photo by and courtesy of Austin Armer.

THE QUICKDRAW

When I was growing up, I wasn't allowed to handle Dad's pistol. Sometimes I would get it out, unload it, strap the holster on, put it in, and practice the quick draw.

So one day when nobody was home, I was in the house quick drawing. When I finished, I got ready to put it away. I put the shells back in it, stuck the pistol back in the holster and made one more quick draw. Boy, that .38-40 made an awful lot of noise, especially in the house!

Well, I heard the chickens outside jump up and raise cain so I thought I might have killed one. I went outside and looked under the tree where they usually were and I found out that I hadn't hit any of them. I just scared the hell out of them.

I never told anybody about it, but I can remember wondering if my dad would see the hole in the wall. It was about as big as my little finger. You might still find that bullet hole but it could be under one of those laths that Dad put on later.

Willis Keys points to the bullet hole visible on the outside of the ranch house which resulted from his childhood "quickdraw."

BILL AND BETTY CAMPBELL

When the Campbells first came out here, they used to come up to the ranch a lot. They were interested in Indian artifacts and liked to talk with the folks to find out places to hunt for artifacts. At first the folks took them out and showed them those places. But when they found out that they were gathering up all of these artifacts and taking them to 29 Palms, they cut them off and tried to keep them from going into a lot of places. Of course, it was open property, but the folks told them that they didn't like the idea of them coming up there taking the stuff out. So after that, their relationship cooled.

I remember the time that Bill Campbell had a crew come up and cut a lot of willow posts from the Desert Queen Wash and the Lost Horse Wash to fence his place. In those days a homesteader out here was allowed to cut enough posts to fence in a certain amount of his property. I don't remember the exact amount, but he was allowed so many posts to be cut off of government land.

Well, my dad found out about it. He felt that Campbell was cutting more than his share and had no right to come all of the way up there to cut them and to take them back to 29 Palms. So he contacted the Department of the Interior and complained and they sent someone out.

Instead of him coming to see my dad, he went to see Bill Campbell. He got into his big car and drove him up to the ranch to talk it over with my dad. Well, the government man said that in this case, they were just going to call it stumpage and let it go at that. So that was the end of that.

In the earlier years the folks took it on themselves to watch out for the whole country up there. If townspeople destroyed Joshua trees or something like that, they'd tell them to stop or they would run them off. They also didn't approve of people coming into the old Indian campground areas and building fires which obliterated the pictographs and petroglyphs. So they often aroused the local people's ire that way.

Bill and Betty Campbell, 1924 homesteaders in 29 Palms, were also early area archaeologists. Photo courtesy of Jack Grover and Paul Smith.

NIGHT VISITORS

The Tucker family was from Tennessee. They used to go back there every couple of years after they got tired of it out here. Then they would come back. They had a couple of homesteads in Twentynine Palms at different times. The original one was off Utah Trail just about a mile from the Twentynine Palms Highway. And then they homesteaded another place down on Mesquite Dry Lake that's within the Marine Base now.

Probably about 1932 the Tucker family which included two boys, a girl, and their mother at that time were staying here. My dad got along well with the boys who were big, husky fellows, and he hired them to help him at the mines doing assessment work.

Nelson Tucker and my dad were over at one of the mines working, while Hugh Tucker was at the ranch. He had been out in the late afternoon hunting for a rabbit or two for supper.

Well, Hugh was coming back just about dusk and down below here at the wash, he heard a funny noise. He couldn't see through the brush too well, and he didn't want to get too close because he didn't know what it was. And he heard some voices, but he couldn't make them out.

So he came running back to the house and got us all together in the front room - his mother, Mary, my mother, and all of us kids.

"I think there's a bunch of Chinese down in the willows!" he said. "I could hear them. They were talking some sort of foreign language!"

At that time there was quite a stir about smuggling Chinese into this country. So everybody got excited and Mrs. Tucker took charge. She asked my mother, "Where does Bill keep his pistol?"

So mother went and got it.

"We'll go down and find out," Mrs. Tucker said.

So she and Hugh took off and hiked down there and found two old prospectors with an old Model "T" sitting around a small bonfire having supper. They were camped for the night. Earlier they had had their dry washer out checking the wash for gold and that "strange sound" was the bellows of the dry washer.

So there was a lot of excitement at the ranch for awhile!

PIANO ROCK

One of the incorrect stories that has circulated about my dad has to do with his involvement with Piano Rock south of the Barker Dam area. There's a big flat rock out there with a whole line of Joshua tree logs in front of it. It's just off the Parks' circular hiking trail that goes through there. That land used to be part of my grandmother's homestead. Dad was going to make a kind of amphitheater there, and he set those logs up there

People used to come up from Palm Springs every year and camp out there. Frank Bogart, who was the mayor down there for many years, used to head the event. It was a big deal.

Well, this wealthy equestrian group would get together down there in Palm Springs and come up here by horseback to camp. All of their gear would come up ahead, and they'd have a chuck wagon where everything was prepared.

While they were here, they'd have catered dinners and a little band for entertainment. Several times they did bring a piano up there and park it on the rock. They had quite a program while they were there, and I guess great parties.

As far as my father playing a piano on the rock, he never did. Sometimes at the ranch he would strum away on the guitar, and he had an old accordion years ago that he used to play a little bit. But he never played the piano.

SMITH—AN ENTERPRISING NEIGHBOR

The area out in front of the canyon leading into Barker Reservoir had been filed on by a fellow by the name of Smith. He came out here about 1929 or '30 and nobody knew anything about him. We later found out that he was a shady character.

He had bought that small cabin from Harmon and moved it over there, and he was building a nice house. He had already finished a big metal shop building which looked huge to me when I was a kid. It was probably 25' x 30' or maybe a little longer with two 10' sliding doors on it. It was big for out here in this country in those days.

Every week or two Smith would have a different car - mostly Model "A"s. Sometimes they were roadsters, sometimes a coupe, sometimes it was a four-door sedan. People began to wonder why he had a different car every so often.

Finally the law came out here and nabbed him. He had been stealing these cars out of state, driving them out here and changing their serial numbers in his metal shop. Then he would peddle them down in Los Angeles. He got a way with this for a year or so until he was caught and taken away.

Well, a guy that knew him by the name of Donaldson took over his 160-acre homestead relinquishment. He was staying down at John Samuelson's place at the time and wanted to move to Arizona or someplace. He really wasn't interested too much in the land and offered the property to my dad for little or nothing. When you buy a relinquishment you are buying the fellow's homestead rights that he had to the property.

After the sale was completed, Dad tore down the house that Smith was building and hauled the lumber home. That was put to good use there when we built that new kitchen He also tore down the metal shop and used most of the metal to enclose the two-stamp mill at the Wall Street.

Dad felt that the change of this relinquishment so many times might not be legal or hold up, so he asked Bill McHaney to file on that 160 acres. And Bill moved in the little cabin that Smith had bought from Harmon and lived there for several years and proved up on the homestead. At the time of his death he had willed it to my dad.

Earlier Dad had my grandmother, Lena Lawton file on a 160-acre homestead over there west of the Barker Reservoir and put a house on it. This was the larger cabin that he had gotten from Harmon. They put it on wheels and moved it intact to her homestead.

My grandmother never lived on the property, but my uncle Al and his buddy, George Legate did and proved up on it. It butted up against the Bill McHaney homestead and with it that gave us a pretty good amount of property and control of the Barker Reservoir.

In fact there were only two small strips of land that prevented the folks from having continuous property between the ranch, my grandmother's homestead, the Bill McHaney homestead, and our Wall Street Mill.

The first was a little strip about a half mile wide from our south gate to my grandmother's property. Dad had been trying to work out a trade with Mr. Whitlow for that strip with some property on the lower end of the ranch that would have tied the ranch to my grandmother's homestead.

The other property was the Bagley homestead between McHaney and the Wall Street Mill. Dad owned over a thousand acres at the time of his death in 1969.

You might wonder why Dad wanted so much land? I don't know whether it had came from the old time when he worked the cattle ranches, and they tried to wire as much as they could for stock grazing. I think that might have been in the back of his mind. In later years I think he wanted to preserve that area through there. He liked it so well, and he hated to see people coming in and raising cain. So I think to preserve it was in the back of his mind too.

Bill McHaney filed on Smith's homestead and lived in the cabin which Smith had moved from Ed Harmon's homestead two miles to the east.

Meyers Dam is partially visible in June 1996 as drought conditions cause Barker Reservoir to dry up.

MEYERS DAM AND BARKER RESERVOIR

One of the consequences of an almost lack of precipitation during 1996 and most of 1997, has been the gradually drying up of the Cow Camp Lake, the Keys Ranch Lake, and the Barker Reservoir, as well as all of the other tanks in the Park. These man-made lakes and tanks were here when the area became a national monument in 1936, and in succeeding years have become water sources for native wildlife.

These "watering holes" owe their origin to the early cattlemen who utilized the desert as a grazing area during the fall, winter and early spring months. The need for water to sustain large herds of cattle was just as important if not more so than an abundant supply of grass and brush on which to graze.

When wells and natural water holes could not be found, the cattlemen created their own by building concrete dams across dry stream beds or washes through which water flowed after rainfall. The resulting pools of water when small were known as tanks, while the largest were lakes. After all cattle grazing in the monument ceased in 1945, the lakes and tanks become sources of water for its native animal population.

During the spring of 1996, the level of the Barker Reservoir was so low that hikers walking across its floor were able to see the top of an earlier dam normally submerged under the reservoir created by the later dam about 50 feet away. By the end of summer the entire dam was exposed to view.

Only a few feet high, this cement and stone wall was constructed with a wooden flood gate. In a letter to the National Park Service in July 1942, Bill Keys wrote that this dam was built by Bill McHaney and George Meyers.

Meyers was a cowboy born in Missouri in 1849 who worked for the McHaney brothers starting in the late 1880's when they were developing their ranch which later became part of Bill Keys' Desert Queen Ranch. Meyers was a witness and possibly a participant in the Frank L. James killing which resulted in the McHaney brothers gaining possession the Desert Queen Mine claim. In a 1966 interview Bill Keys said that Meyers received cattle from McHaney for his part in the shooting.

When the McHaney brothers sold the mine in September 1895 to a Denver firm, George Meyers filed suit against them for one third of $10,000 which was his share of the ore that had been milled.

Meyers continued to run cattle in the area and on December 19, 1903 went into partnership with C.O. Barker of Banning. By this agreement Barker furnished his interest and title to his buildings and fences, his Desert Queen Stock Ranch at Cow Camp, his watering holes, and 115 heifers and 7 bulls " branded C.O. on right hip." In return Meyers would take care of them year round providing them with feed and protection, as well as to "keep in good repair all buildings, fences and improvements returning them in as good condition as when they were received." This lease was to last until January 1, 1908 when the original number of 122 animals would be returned to Barker and the increase split.

In a letter from Meyers to Barker in February 1904, Meyers wrote that he and an Indian cowboy working with him had found 11 dead cows at Warren's Well (today's Yucca Valley), that the well was all right, that

Meyers Dam in foreground and Barker Dam in background as the reservoir continues to drop one month later in July 1996.

Meyers Dam is completely exposed as Barker Reservoir is dry in November 1996.

the grass was good at the ranch and after the cattle were moved there, he would work on the well. He also requested that the following supplies be sent out to "Worns Well: 2 plug tobaco, 1 sack bens, 1 can baken powdr,1 can lard, and 1 galen coloil." He wasn't a good speller!

Meyers dissolved the partnership sometime in 1906 realizing that the desert was only good for cattle in winter and not year-round. Barker hired other cowboys who operated out of the Desert Queen Ranch's Cow Camp and until 1913 they drove the cattle to Banning in the spring, returning to the desert in late fall.

One of the lasting results of Barker's involvement with Meyers was the building of a larger dam downstream from the Meyers dam about 1902 when the land around the small lake was public domain. This structure measured about 150 feet long by 9 feet high.

In 1913 Barker went into partnership with Will Shay, brother of Walter Shay, who served as sheriff of San Bernardino as did his son and nephew. The Barker-Shay partnership took out more water leases and by 1915 they had 25. Ten years later by the time the partnership dissolved, they controlled 50 water sources. Because of this, Barker and Shay were able to effectively increase their grazing area across the High Desert to about 700 square miles.

On December 12, 1914, the Barker Reservoir and surrounding area was withdrawn from the public domain and designated as Public Water Reserve # 14 which allowed anyone to use the water created by the dam.

In 1917 when Bill Keys filed for a 160 acre homestead on the old Desert Queen Mill site, it included the Cow Camp area and buildings used by Barker and Meyers and Barker and Shay. This resulted in a number of conflicts between Keys and the cattlemen in the years that followed. The use of Barker Reservoir also became a source of contention between Keys and local cattlemen, and later with the National Park Service.

Willis Keys said that the second dam built by Barker and Meyers was "rather poorly constructed." It was made with laid-up rock which was filled in with sand and dirt. Then both sides were plastered. It leaked from the start and wouldn't hold water very long. Bill Keys, who used it to water his own cattle, decided to make some needed improvements of his own.

In 1932 he went inside the dam, dug down to bedrock, and started a new one. He poured a wider concrete base about two feet at the bottom right against the old dam. Then he raised it higher than the Barker one. In 1938 he added additional height, and in 1949 and 1950 it was increased to its present one. When full, the lake covers nearly six surface acres. About 25 feet below the dam to the west was a low wall that impounded leakage from above.

In September 1932 in an attempt to control the water impounded in the Barker Reservoir, Bill Keys located a millsite claim that included dam and the water behind it. The following year he had Bill McHaney file on a 160-acre homestead that he had purchased from a man by the name of Smith who had failed to prove up on the property. Because this land controlled the south entrance to Barker Reservoir, Keys had friend, Bill McHaney file on the property and will it to him at his death.

Lena Lawton, mother of Frances Keys, controlled the western approach with her 160-acre homestead. The north side of the lake was bordered by high rocks and was virtually unaccessible. With the dam and

Bill Keys working on the extension of Barker Dam in 1938.

lake surrounded, Bill Keys used it for his cattle and threatened to deny access to the public.

At the base of the dam, Bill and Willis built a double-ring watering trough in June 1939 marking the date in the cement. Water from the lower catch basin was piped into the trough where a float valve in the center regulated the flow. The inner rock ring protected the device from damage by their cattle. This concrete and rock trough replaced the wooden box trough that still remains nearby.

Bill's closing off of the Barker Reservoir also caused conflicts with officials of the National Park Service after the monument's creation. His Big Chief Millsite was declared null and void because the land was part of a water reserve.

In a July 1942 letter to the NPS Regional Director in San Francisco, Bill emphasized that he had been using the water for the past 30 years for mining and cattle watering purposes, and that "it has never been used by the general public. In fact it is stagnant rainwater and is not fit for any other use than what I used it for."

In 1943 The Park Service conceded that "there was no way of stock reaching the dam without crossing land owned or controlled by Mr. Keys," and because "he had watered his cattle there for so many years, we would be hard put to deny him his right now that the water reserve has been superseded by the National Monument."

In May that year after Worth Bagley ambushed Bill Keys and was subsequently killed, the Park Service cancelled Bill's grazing permit. While this put an end to the conflict over cattle grazing at Barker Reservoir, in 1949 and 1950 Bill Keys assisted by his wife, Frances and children, Willis and Phyllis poured additional concrete to the top of the dam bringing it up to its final height further increasing the lake's capacity.

The Park Service evidently were unaware of or chose not to interfere with the renewed work on the dam, as a review of both year's Superintendent's Reports makes no mention of this building activity. The only mention of Mr. Keys in those reports was in February 1950. "After observing four horses grazing in Lost Horse Valley, a letter was sent to homesteader W.F. Keys advising that this was contrary to rules and regulations. The horses have not been observed since the letter was sent."

In September 1997 after more than a year and a half of drought, heavy rainfall fell on this desert location and the subsequent runoff once again filled much of the Barker Reservoir, replenishing a source of water for Joshua Tree National Park's native animals. Another result of the rainfall was the disappearance of the Meyers' Dam under the dark waters behind Barker Dam.

Double-ring water trough built by Bill & Willis Keys in June 1939.

Willis and Phyllis Keys working with their dad on the last extension of Barker Dam in 1950.

JOHN SAMUELSON

Erle Stanley Gardner first met John Samuelson on February 9th, 1928. It was a cold blustery day, and he had moved his camp wagon down into the shelter of one of the big granite walls to get out of the wind. At Quail Springs where he stopped to get water, he encountered Samuelson, who was there for the same purpose. In the course of several trips to the desert, Gardner learned more about John Samuelson and wrote of him in *ARGOSY* magazine and in his book, *NEIGHBORHOOD FRONTIERS*.

John Samuelson was born in Sweden in 1873 and said that his early life had been spent at sea. This five foot, eight inch tall man appeared at Bill Keys' ranch in 1926 looking for work. Keys hired him to help with his Hidden Gold Mine located below the overlook where Keys' View is today.

By 1927 Samuelson had decided to homestead and located a piece of property in the valley southeast of Quail Springs. Here on the top of a small hill, he build a wood and canvas shack and began to carve political sayings on the rocks that surrounded it.

After he filed on this homestead in 1928, the land office discovered that he was not a U.S. citizen and ruled that he could not legally hold title to the land. As a result, the Headington family homesteaded his land, after he moved to the Los Angeles area. Samuelson returned to the desert from time to time during the next ten years to continue his mining pursuits.

On Christmas Day 1942 while at a dance in city of Compton where he was living, the then 69 year old Samuelson got into a heated argument with two men. He shot and killed Antone Grassman and wounded Cecil Wright in the arm.

Years later when Erle Stanley Gardner began looking for Samuelson, he discovered that he had never been tried for the murder or shooting. After his arrest and prior to his trial, Samuelson had been adjudged insane and sent to California's State Hospital at Mendocino. When surveillance over him was relaxed, he made his escape, eventually going northward to Washington.

It was from here that Bill Keys received his first letter from him, and in 1954 he received his second and last letter. In it Samuelson wrote that he wanted to return to the desert but was afraid that he would be caught by the authorities.

Not long afterwards, Keys received a letter from officials at a logging camp where Samuelson was working. They wrote that he was in serious condition having been injured in a logging accident. Soon after, another letter came informing him of his old friend's death from those injuries.

Samuelson's wood and canvas house burned down in the 1930's after his departure from the desert, but the eight, flat-faced rocks with his inscriptions can still be seen today ncar Quail Springs by hiking a mile and a half from the main road that leads to the west entrance of Joshua Tree National Park. While some of the words are misspelled due to his poor command of the English language, his political views are as clear today as they were when he carved them seventy years ago.

Two of the rock inscriptions left by John Samuelson in Lost Horse Valley.

JOHNNY LANG'S RANCH VISIT

Johnny Lang walked whenever he came to the ranch. He had lost his burros and mules, so he walked everywhere he went. I can just barely remember him coming to the ranch, and it was probably the only time that I ever remember actually seeing him alive.

My mother often told the story about one of Johnny's visit. She was in the main part of the house. The old kitchen was out in front then, and there was a screened porch in between it and the house. We used to keep our meat wrapped up sometimes there on the kitchen floor, and the meat saw hung from a nail.

Well, on this one visit, Johnny came in to the ranch and nobody was around so he walked across the porch into the kitchen. And the first thing that he did was to go over to the meat saw and run his fingers down it. He wanted to see if there was any fresh meat in the house.

Johnny Lang about 1923.

JOHN LANG

On a cold day in January, the old miner in need of supplies tacked a note to his cabin door that he was heading out and was expected to return "soon."

Two months later his mummified remains were discovered amongst the desert brush by Bill Keys, Frank Kiler, and Jeff Peeden as they were constructing the Keys View Road in what is now Joshua Tree National Park. The body of Johnny Lang was still covered with a piece of canvas, while nearby the remains of a small piece of bacon wrapped in paper, and the ashes of a last campfire were silent witnesses to what happened that night. Lang's advanced age, his state of malnutrition, and a cold winter's night combined to make this camping spot his final one.

Keys was instructed by county authorities to bury Lang where he had been found. With Jeff Peeden standing by with a shovel in hand, Bill Keys read a service over the remains of his long-time friend, and seventeen year old Frank Kiler took several photographs to record this desert burial on March 25, 1926.

*   *   *   *

John Lang was born in Missouri in 1853, the son of George Washington Lang, a native of New York.

Both John and his father were cattlemen who had moved west to Kansas, then later to Texas, and on to Arizona finally arriving in 1889 in Los Angeles where they determined it would be an important market for their cattle.

Soon the Langs and a number of Arizona cattlemen bought grazing concessions in Mexico and became involved in an enterprise of driving cattle from Mexico into Arizona and then across the Colorado Desert to market in Los Angeles. They figured they could save $4000 or more over the freight cost of shipping a thousand head by train.

In 1891 the Colorado River had spilled over its banks creating new grazing pastures south of Indio, and by 1893 the Langs had established their headquarters in the Indio area. It was during this time that one of their horses wandered away into the neighboring mountains of what is now Joshua Tree National Park. While trailing his mount, John supposedly looked down and picked up a piece of ore that was so rich that even he as a non-miner recognized it as gold.

Other miners in the area had discovered the same rich hillside, so on January 3, 1893 John took a copy of the claim notice into Riverside to be recorded in the names of John and George Lang, Ed Holland, and James J. Fife.

Bill Keys digs the grave for Johnny Lang whose mumified body lies in the foreground.
Photo by and courtesy of Frank Kiler.

Jeff Peeden stands while Bill Keys says a few words over the open grave of Johnny Lang,
March 25, 1926. Photo by and courtesy of Frank Kiler.

The partnership formed the Lost Horse Mining Company and the first year the ore was producing up to 25 ounces of gold per ton—about $3500. Beginning in 1895 the partners began selling off their shares due to the accusations that John Lang was skimming off the profits while transporting the gold-mercury amalgam to Los Angeles.

By 1896 George Lang had sold his shares to James and Thomas Ryan leaving John with a quarter interest in the mine. The Ryans ran the day shift, and John worked the night. A noticeable difference in the amount of amalgam produced finally caused the Ryans to hire a detective to watch Lang and confirm that he was taking part of the amalgam from the night shift. The Ryans offered him the choice of buying them out, selling, or going to jail. John Lang sold his share to the Ryans for $12,000.

Lang continued to live in the area and to prospect further to the west. Stories circulated in the area that he supplemented his prospecting finds with the amalgam that he had skimmed off of the Lost Horse Mine in earlier days. That mine was operated by the Ryans and others up until 1923, afterwhich it lay idle until 1931. It was during this inactive period of the Lost Horse Mine that Johnny Lang was living part time at one of the mine's cabins. And it was from here that he departed in January 1926 to get some food and supplies, fell ill two miles away and died.

Two months later his friend Bill Keys, while building a dirt road to his Hidden Gold Mine, found Johnny's body and subsequently buried him. At the time of his death, John Lang was 72 years old, 5' 8" tall, balding, light complexioned, with brown hair and blue eyes.

Frank Kiler was 17 years old when he was spending his Easter vacation at the Keys Ranch and took the photographs of Johnny Lang's burial. On a return visit to the ranch in Nov. 1996, he and his long-time friend Willis Keys talked over old times.

A FEW TRUCKS, A CAR,
A TRACTOR, AND A JEEP

Dad had at different times a collection of old automobiles and trucks. He didn't acquire the old 1922 model Mack truck until later years, and he only used it a little bit. Originally it was a dump truck owned by the City of San Bernardino. It probably went through a number of hands before my dad acquired it.

It really wasn't a practical truck for out here in the desert, because if you got into one of the sand washes, it had very poor traction with its solid rubber tires. So Dad never used it too much. It was also rough riding and hard steering, so he didn't like to drive it. He mainly drove it just around here He'd rather drive the old Maxwell truck which he used mostly until he got his jeep.

I resurrected the Mack when we were out here in 1948 mining at the Desert Queen Mine. We loaded an air compressor and an engine on it and hauled it up there. We went down that back road into the mine and over there next to the tunnel. And we ran the air compressor with this old truck. We jacked up the rear wheel, put it in gear, and had a belt running to the air compressor to run it. On that trip Dad and I got down into that wash at the bottom of the hill and got stuck. The only way we got out was with a chain block and a chain wrapped around a big boulder on the other side. Two of us chain blocked it across that canyon.

1922 Mack truck used at the ranch and driven by Bill Keys in the Disney film "Wild Burro of the West" in 1960.

1914-15 Traffic truck with the unfinished framework that Willis was building to help separate sand and gravel for making concrete.

Dad wanted me to help him get it back out to the ranch, so we worked on the road a little bit. Then I took off with it and headed up that hill. Dad was ahead of me, I guess, pulling. And he pulled me and I came right up that hill. The hard tires didn't get very good traction and I remember it being hard to steer. Dad did drive it though when they were filming "Wild Burro of the West" out here in 1960.

Near the Mack truck at the ranch is the remains of another truck. The old frame was a Traffic truck that dates from the early 1900s and was brought out here in the early 1920's or maybe a little before by the Gold Coin Mining Company to haul ore. A fellow by the name of Rand had that mine then, and he brought that truck out. It had the same problems as that old Mack with its. hard rubber tires. My dad said that you could see that rig along the road stuck in the sand at almost any time.

My dad acquired it years later around 1940, and he just bought it for junk. It had been stripped down pretty well. When he got it, it still had the motor in it - an old 4-cylinder and a cast-iron radiator, but everything else was gone. I guess that it had sat down there near the Gold Coin for a good many years.

That cylinder up front was the gas tank just like on the old Maxwell. This Traffic had a peculiar drive system. A lot of the old trucks were chain drive, but this one had what they called an internal gear drive. It had a small differential setup and did the gearing out here in the wheels. It has a little pinyon gear that drives an internal gear which is a little different than most.

This truck also got stuck so often that we never did a lot of hauling with it. I think it was in the early '50s that I decided to go into the sand and gravel business. There's a lot of sand down in the lower wash, so I thought I would screen that sand and sell it to the Marine Base and to builders in Joshua Tree and 29 Palms who were wanting good concrete sand. But in order to use the sand, it had to be screened.

Dad had several big rotary screens to use. So I built that framework to mount the screen, then I could load it with a skip-loader with sand, run it through the screen, and remove the debris. But the framework was as far as I got, and I never got it finished. I had too many other projects going trying to make a living with a bulldozer and truck.

We also had another vehicle out here that was somewhat older. It was brought out in 1910 or '12 by the people who had the Black Warrior Mine. They had it especially built by the Chase Company to use out there to haul ore. Originally it had a 3-cylinder, air-cooled, 2-cycle engine, and it was a quite piece of equipment even in later years. My dad acquired it when the mining company went defunct. The old engine was pretty poor, so he put an old 4-cylinder Dodge engine in it, fixed it up, and used it for many years around here to haul wood and ore. It had wooden wheels with iron tires and a chain drive.

He used to take it to the Cherry Festival in Cherry Valley and Beaumont and run it in the parade every year or two. On one of these trips, it burned out a rod or bearing, so he parked it and left it at a garage down there. He didn't get back for many months, and by then the owner had placed an $800 lien on the vehicle. He couldn't afford to pay that money for the old truck and was never able to get it back. I tried to get it back once but couldn't either, so that's how it got away from us.

Another truck we had at the ranch was originally an armored car. It had belonged to a fellow down in Joshua Tree who was a well driller and had a well rig. He wanted something that he could use to pull his rig with out in the desert. So he went to a government surplus auction up in Port Hueneme and bought this old military armored car. He stripped off all of the iron and put a fifth wheel on the back to pull his well rig. He also mounted an arch welder on it and used it for quite awhile down there.

Finally he gave up well drilling and parked it in his back yard. I had admired it for a long time, so one day I asked him what he would take for it. I think he said about $250, so Dad helped me finance it and we went down and bought it in 1953 I believe. Dad

This truck was once an armored car that Bill Keys adapted to ranch use by adding a bed. Photo by and courtesy of Mel Weinstein.

put a bed on it, and it was quite a vehicle. It still had the extra plating, springs, and heavier equipment on it to support the drilling rig. I also remember it only had four or five thousand miles on it. It would go just about anywhere. i don't think we ever hauled wood with it. We mostly used the old Maxwell truck for that.

My dad bought out a junk yard once and moved all the stuff up here to the ranch. He had piles and piles of iron and stuff that wouldn't quit. It came in handy with all of the improvising he had to do especially mining and running the mill. Why you just couldn't go down to the corner and get a part. If you didn't have a part or weren't able to make it, you were in bad shape. But we did it all. If we didn't have it, we made it out of something.

I believe that originally the Fordson tractor was brought out here probably in the 1920's to be used at one of the mines over there in Gold Park area. Phil Sullivan who used to live in Twentynine Palms was associated with this outfit at one time, and he acquired the tractor eventually, and my dad acquired it from him.

It had a framework and a large hoist out in front, and they'd use it up there in that Gold Park area for hoisting ore from one of the mines. So when my dad got it, he brought it up here and took the hoist off, I remember. And he sold that, for he had no use for it, but he wanted the tractor.

He did do quite a bit of ploughing and used the Fordson tractor to plough in the orchard area and also out there where he planted the wheat. There's an old disc plough out there that we used to pull it with for ploughing. He also used it for running the old saw for cutting firewood. And that was about the extent of the use of it around here.

Fordson tractor with a dented radiator caused by the family's billy goat.

Remains of the 1928 Dodge that Bill Keys was driving when he was shot at by neighbor Worth Bagley on May 11, 1943.

If you look at the radiator closely, you can see it's been bashed in on one corner. We had a mean old billy goat who had big horns. You couldn't get out in the yard before he would take a run at you. So my dad used to tie him, or did at one time, to the tractor. And by golly, the old goat decided he didn't like being tied, so he bashed in that radiator with his horns.

At one time my dad had this tractor completely apart to overhaul it. We had a coil for each cylinder, and a magneto created the electricity to run it. The main wire came right off it here and went back in here to another one of those posts. It had what they call a timer. We call them a distributor now, but that's more or less the same thing. There was a coil for each cylinder, and when you were turning it over you could hear these things buzz. They vibrated each one when it was ready to fire. And they gave a good hot spark. I found out. When you get tangled up in one of those, it will really jolted you.

The car Dad was driving when he was shot at by Worth Bagley was a '28 Dodge. It was called a Victory 6 or Dictator 6. George Trinkle, friends of my parents, gave Dad this old car. You can still see where Bagley's bullet struck the edge of the door on the driver's side.

My dad had started to move the Joshua tree logs that Bagley had placed out of the road when he saw Bagley coming toward him. So he ran back to his car. The door was open and my dad reached in to get his rifle off the seat of the car when Bagley shot at him. And whether the bullet went in front of my dad or behind him, I don't know. It came pretty close.

The car had a soft top. It was made out of a rubberized cloth material that looked kind of like black vinyl. There were cross bars and slats to support it.

Dad had two of these '28s. One that J.D. Ryan gave him which had belonged to his son, Sam who had passed away. It had sat over there at the Lost Horse Well for quite awhile. Dad asked him what he was going to do with it, and Ryan asked him if he wanted it. So Dad drove that around here for a long time. Then George Trinkle gave him this one, and he drove it for quite a few years.

Dad's jeep was his favorite vehicle. He had traded a mining claim to a doctor for that jeep in '48. With its four-wheel drive he could go just about any place that he wanted to go. So he put a lot of miles on it and drove it around for many years. He got a lot of use out of it.

It had a canvas top on it originally, and of course it wore out here in the sun. So I said, "Dad, let me have that thing for a few days and I'll make a top for you." That was right after Gwyn and I got married in '56. So we took it down there to Joshua Tree where I found an old Model A touring car body. I had an acetylene torch so I got busy, cut and fitted it, then put the hard top on. There was a mix up when the government bought the ranch furnishings after Dad died. The jeep wasn't part of the deal, but I was never able to get it back. I still have the pink slip on it.

Bill Keys' 1948 Jeep that he drove until a few days before his death in June 1969.

The body of an old Ford protected ranch chickens from coyotes, served as a roosting place and had a nesting box in the truck. Photo by and courtesy of Mel Weinstein.

MODEL T CONVERSION

I was always fascinated by the machinery that my dad had out there in the yard. You can't realize today how much there actually was at the ranch in the early days. I would go out and look at automobile parts and turn them and try to figure out how they worked. I was probably about five or six when I really got interested.

I'd go help myself to my dad's tools. I usually left them laying out there somewhere, and I'd get my butt paddled for not getting them back to where they should have been. But I tinkered around with a little bit of everything.

When I was probably about ten years old, my dad gave me a Model T. He said, "Go ahead, if you want to fix it up." It was about a 1924 I think, and a two-door sedan. It was a pretty nice old car.

The fellow who had owned it was a prospector. He was coming out of Pleasant Valley heading up towards the ranch when he burned out a rod down there. So he walked in and told my dad what had happened.

He said, " I'd like to get a couple sticks of dynamite and go back down there and just blow that thing up!"

And my dad said, Oh, you don't want to do that! We'll go down there and tow it in."

So they went down and towed it in and parked it down there by one of the old tent houses. Johnny Ulman was his name, and he stayed for a few days and then left. He never did show up again.

So the car sat there for a year or so. And that's when I kept bugging my dad that I wanted to work on it. So he said, "All right, go ahead!" We pulled it up there by the barn, and I took the engine apart.

My dad had all of these Model T parts, so I found a rod which I needed to put it back together. We didn't have any gasket sets, but my dad had gasket material, and whenever he needed a gasket, we'd just cut our own after we had made a pattern. Sometimes I just used thin cardboard for gaskets depending upon what it was for.

So I got the old car running, and it ran pretty good. The tires were very poor on it, but I located enough old tires that I could cut the beads out of some of them and put them over the tops of the others and get by that way.

The bead is where the tire goes around the rim. It's reinforced with metal wire. So if you cut that bead out, then it makes the rest of the tire more pliable, and I could skin it over the top of the other tire. So I ran on these tires for quite awhile, but I had to do quite a bit of repair work because they were pretty poor.

My dad used to get gas whenever he went in to 29 Palms, Banning, or Indio. He had a couple of tanks that would fit in the back of a car. One held 18 gallons, and one held 20 gallons. He would get those full of gas, and that was our usual supply. You had to treat gas sparingly, so I got rationed a gallon at a time I think to run around the ranch with that old car. When my dad wanted me to take him to the mill or mine or Pleasant Valley or someplace else, then he would dish out a little more gas. I even took my mother down to 29 a few times in it.

Then I decided that I wanted to make a cut-down car out of it. I didn't like the big, tall, sedan, so I took the body off.

My dad had an old home-made roadster body, and I decided that I would use that. But before I put the body on, I thought I'd like to have a gear shift. The old Model T just had three pedals. All the way out on the left one was high gear. Half way in was the clutch, and that put it out of gear. All the way down to the bottom was low gear. So you'd just push all the way down and take off, then let it back through neutral into high gear. That was it - two speeds.

The center pedal was reverse. When you wanted to go into reverse, you held the left pedal half way down so the clutch was out, then hit the reverse which was the middle pedal. Then you could back up. The right pedal was the brake.

Well, I liked the gear shift deal which was a little more modern, so I tried to figure out a way of putting a gear shift and transmission in that car.

My dad had an old Chevrolet transmission. I got it out there and hunted for parts to make the hookup for the drive line and the U-joints. Soon I found enough that fit, but then came the question of the length. I had no way to cut the drive shaft and to reweld it. Instead I cut the frame and lengthened it to the length of the transmission.

Then I spliced in angle iron and bolted the frame back together. So it was actually a foot longer than it was supposed to be. I also used an Overland radiator instead of the Model T one. This made it an odd-looking thing, because the radiator was quite a bit higher than the body. Someone had given me a Whitfield carburetor for a Model T and I put that on. There wasn't any windshield either, but this car ran.

So I got it going one day and was headed down the road out of the ranch. I had my foot all the way to the floor board and really got it going along pretty good. Then I saw a car coming up ahead of me out of the wash.

"I've got to stop!" I thought. So I jerked my foot off the throttle, and I had to bring it around the gear shift to get to the brake, and I knocked it out of gear.

When that transmission was out of gear, you didn't have any brakes up there in the Model T part of it. There were no brakes at all. So I tried to turn out of the deep ruts which were about a foot deep. Well, that thing didn't want to turn. I just leaned on it as hard as I could, and this car was still coming.

Just at the last second, it caught a little bit and turned out. I did a big brodie out there, and dust flew up everywhere. The people stopped and wanted to know if I was all right. "Yes, I was fine."

After that experience, I ran that car around quite a bit. I even got my mother to ride in it a few times.

Bill, Virginia, and Phyllis Keys riding in their 1910 Chase vehicle in a Cherry Festival parade.

A TRAGIC ACCIDENT

It was in the summer of 1935 right before my first year in high school. This young fellow who I think was about eighteen or nineteen had been coming out here for several years to spend his vacations. His name was Bob Schaffer and he was a nice kid. He was going to a military academy, and his folks were pretty well off.

In the summer he'd come out here for a couple of months and camp in one of the little cabins at the ranch. He would just roam around, and he liked to play army. He had a place way back up in the rocks where he set up a little military encampment. He had these little lead toy soldiers and he would set up a whole battle field with those. Then he would get up there with his .22 rifle and shoot from one side and then from the other. He was really enthusiastic about that He also used to take black powder and make some little bombs.

So this one day he asked me, "Could you get a stick of dynamite, and we'll go over somewhere and set it off?"

And I said "Yes. I know where the dynamite is."

So we went and got a stick a dynamite and then went up in the junk pile and found a cast-iron hub out of an old wagon wheel.

"That's about the right size," he said.

So we got into his car and went over to the Barker Reservoir and found a place in the rocks to set it off. He loaded that dynamite in that cast-iron hub and put the fuse in. There was a crack under the rock with just a little shelf underneath, and he put it under there.

My brother, Ellsworth, was with us and we all ran up on top of some rocks about a couple hundred feet away.

And I said, "I think we ought to get back further."

And Bob said, "Oh, no. That rock will absorb the blast."

So we sat there, and it didn't go off.

About that time three other fellows who had been out hiking came along. It was Bob Craven, a doctor's son and a couple of friends of his. They came along and asked what we were doing. "We've got a bomb down here." So they decided to stop and see what would happen.

So Bob Schaffer went back down and found that the fuse had gone out. He relit it, came back, and sat down between my brother and me.

We were all sitting on this sloping rock waiting when it went off. I heard Bob just sort of sigh and I looked, and he was just starting to fall forward. So I reached out and pulled him back. He didn't make a sound except for that little grunt.

At first I couldn't figure out what had happened. Then I saw the blood coming out of his chest. A piece of that wheel hub had hit him and gone right through his heart. It had killed him like that.

When that thing went off, I could hear that schrapnel hitting all around us. So it is a wonder that the rest of us weren't hit too!

Well, Bob Craven and I picked him up and packed him down out of the rocks and loaded him in the back seat of his four-door '34 Plymouth. I drove his car and Bob Craven had his own. I parked Schaffer's car at the crossroads at the big wash and rode on with Bob to tell my folks. Someone at the ranch went down to Yucca Valley to phone his parents and the coroner. It was a real shock!

Former tent house that served both as a schoolhouse for teacher Oran Booth and pupils, Willis and Virginia Keys, and later as a bedroom when Willis returned to the ranch from attending high school in Ontario.

SCHOOL DAYS

The first teacher at the ranch was my mother, who taught us a lot. When I got around ten or eleven years old, my mother was worried about our schooling. She'd sit down with me and my oldest sister, Virgie, and would try to teach us how to spell and to read. We were of school age but there were no schools around. It was too far to go to San Bernardino. So after breakfast and after the chores were done, we would sit down in the house for two hours or so and have school, and that would be about it for the rest of the day. Of course, her time was limited because she had so many chores to do that she couldn't spend a whole lot of time doing it.

Different people who were either working for my dad or coming up here to stay for awhile would also teach. Most of them would come up and spend a half a day with my sister and me.

I remember a fellow who came to stay. He was a World War I vet, and I think he had been gassed. He had a little pension and wanted to stay out in the desert. So he stayed in one of our tent houses and boarded there at the ranch. Mother asked him if he could do a little teaching, and he said he could. So I think we went up there to his cabin a couple hours a day. He wasn't too sharp as I recall and liked to draw cartoons mostly. He would give us a little instruction then draw a few cartoons. Then he'd have us draw a few.

Then the first real teacher who was accredited taught us for about six months in the little cabin up in the corner of the ranch. Mr. Oran Booth actually had taken teacher's training and had been a teacher--a shop instructor at some school at one time. He worked for my dad for awhile, and so he'd agreed to take on the chore of trying to teach us.

Our little school house which started out as a tent house was used as a guest house. In years past people used to come up here, and they would want to stay overnight or for a couple of days. So my dad said they he would build some little cabins to rent.

He had several old tents from the mining days. So he built wood floors with wood sides and wooden framework for the top. On them he put these tents.

But renting them never worked out too well. There wasn't too much travel in here. Mostly when people came, they were friends, and they didn't pay. And that's the way it was out here in the early days. People dropped by, and some stayed a month or just overnight and for breakfast. That was the normal way.

Well, after two or three years the tents went to pot. Out here in the sun they didn't last long. So he just finished them in with lumber. One of the four cabins was used for our schoolhouse at one time.

When we went to school here, we just had a table and a couple of chairs. The school desks here now were never in it when we had school here. These desks were used in the last school house south of the main house. They were moved out of there for some reason and brought up and stored here.

After Mr. Booth left, it was another year before my dad finally hired Lela Carlson, who was from Alhambra. She had originally taught out in the desert at Ludlow that year or two years before. She had come through the ranch with some of her friends one time, stopped and talked with the folks and was interested in teaching here. She and my folks came to an agreement that she would be paid $30 a month and room and board. That was about 1934.

She taught in the cabin by the lake. My dad had originally built it for a guest house in the teens sometime. The folks used to rent it out occasionally, and for many years this was the only cabin for guests. When we got Miss Carlson to come up to teach, he fixed it up for a school.

When we had school, classes were usually held inside because most of the time we were here in the winter and it was cold. We had that little old wood stove in the main room, and we'd stoke it up and sit around and have our classes. My sister and I were the only ones going to school then. There were two beds there and in other times there was a bed here. When the weather was good the teacher slept out on the porch.

Lake cabin used as the teacher's residence and classroom during the 1930s was stabilized and restored by the National Park Service in 1993.

109

Running water in the kitchen was piped in from a outside tank that sat back up on the rocks. But it was too big a chore to keep the tank filled, because we had no means of pumping it at that time. Usually, I think we'd just bring the teacher a couple buckets of water a day.

Before Dad built the dams in here, this used to be our main way to go up and look at the lake. We would just go right up the rocks to a nice little flat place along side the lake where we could have a picnic or swim. But with the dam there, it made it impossible to get through there. Other times we'd go down in front of the cabin and play a little baseball.

Lela Carlson taught my sister, Virgie, me, and I think my brother, Ellsworth went part-time also. I was thirteen then and had a lot of ground to make up.

The subjects that we were taught in the school were mostly spelling, arithmetic, and reading - the bare fundamentals. And of course, a little bit later as we learned to read, we went into history, grammar and composition of the English language and things like that. We had quite a bit of reading on history.

I finished up my eight grades in a year and graduated in 1935. So you can tell that the teacher was really on the ball. Burton Thrall, later the County School Superintendent, came out from San Bernardino and handed me my diploma. We had a little ceremony there in the teacher's cabin.

That fall I started high school at Chaffey in Ontario. Lela was followed by Mrs. Marsh, Miss Starr, and then the Dudleys.

When I went to Ontario my folks brought me in and stopped at Warren and Daisy Kiler's to ask them if they knew anyone who would be interested in boarding me. My folks were hoping that they would board me, but both boys, Frank and Delmer, were living at home then, and they didn't have any room. Warren thought that maybe the people who were renting next door would be interested, and they were.

The Willages had two daughters and a son. The father, son and one daughter lived in Los Angeles during the week and came home on weekends. The father and son worked over there, and one daughter went to Aimee Simple McPherson's Bible College. The other daughter worked in Ontario as a beauty operator. So the first year I lived with them until they moved to Los Angeles.

The next summer, the folks were worried who I would board with. So we went back in, and Mrs. Kiler said, "There's a new couple over there in that rental. They're older and might be interested."

So we went over and talked to the Willets. They were in their mid-sixties or maybe a little older. He had been a stern wheeler captain on the Mississippi River for a good part of his life, and after he retired they had moved to California. They said, "Sure. We'll take him in."

Mr. Willet was deaf, and I wrote everything that I wanted to communicate to him. We got along fine. They were a nice couple, and I lived with them for the next three years until I graduated in 1939.

During my four years at Chaffey, I managed pretty well. It was a little tough in some of the subjects because I didn't have quite enough background that you acquire over a period of four or five years of going to grammar school. Otherwise it didn't bother me, and I left high school in pretty good shape.

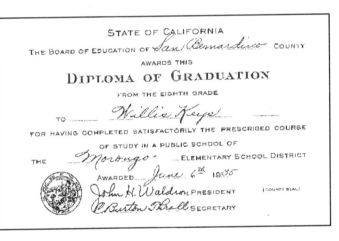

STATE OF CALIFORNIA

THE BOARD OF EDUCATION OF San Bernardino COUNTY

AWARDS THIS

DIPLOMA OF GRADUATION

FROM THE EIGHTH GRADE

TO Willis Keys

FOR HAVING COMPLETED SATISFACTORILY THE PRESCRIBED COURSE

OF STUDY IN A PUBLIC SCHOOL OF

THE Morongo ELEMENTARY SCHOOL DISTRICT

AWARDED June 6th 1935

John H. Waldron PRESIDENT (COUNTY SEAL)

C. Burton Thrall SECRETARY

Above: Willis' diploma from the eighth grade at Keys Ranch. C. Burton Thrall, later the Superintendent of San Bernardino County Schools, came to the ranch to personally present the diploma to him.

Right: Willis Keys' graduation picture from Ontario's Chaffey High School. He graduated in June 1939 "in pretty good shape."

ORAN BOOTH
Teacher, Miner, and Friend

Oran Booth was the first professional teacher to come to the Desert Queen Ranch and to hold regular classes for the Keys' children. Willis and Virginia were his pupils when he taught there in return for room and board starting in 1930.

He had been born in Truckee, California in 1901, the eldest of five children. Truckee was a railroad town at that time, and Oran's father worked for the Southern Pacific Railroad as a brakeman and later as a conductor. As his father's railroad assignments changed, so did the family's location, first to Rockland, and then a year later in 1908 to Roseville.

After the death of his father in January 1920, Oran moved to Los Angeles where he worked briefly before heading back north to Summit to work for the railroad on a shed gang. His job was to help with the rebuilding of the wooden sheds that protected the train tracks from heavy snow falls at the higher elevations. He also laid tracks, and after a few months was assigned the job of helping the three Chinese cooks with cutting firewood, keeping the water barrel filled, burying the garbage, and cleaning the railroad cars used for the workmen's sleeping quarters. His free time was spent canoeing one of the nearby mountain lakes and riding his motorcycle.

His mechanical expertise at keeping his motorcycle running came to the attention of a co-worker whose father managed an auto supply company in Sacramento. This led to a job offer there where he became a specialist in fuels and carburetors. A year later he and a friend opened their own carburetor business in San Francisco, then moved on to Los Angeles to work for several similar companies there.

In the fall of 1922 he returned to Roseville to complete his high school education while he again worked for the railroad. After graduation he returned to Los Angeles and worked until 1924 when he entered college at Santa Barbara. Oran graduated in 1927 with a B.A. in education and accepted a teaching job at Woodrow Wilson Junior High in Pasadena.

His interest in teaching waned after one year, and he became interested in getting back outdoors and into mining. He resigned his position during the summer and came to the Twentynine Palms area in 1928, and with partner, Earl McGinnis, filed on a millsite in what is now Joshua National Park. They called it the Wall Street Mill, because Earl thought it sounded "like a lot of money."

Oran dug out and re-established the well that had originally been dug there by Bill McHaney in 1898, and began prospecting in the surrounding

Oran Booth in April 1979 on a return visit to the ruins of the Wall Street Mill cabin that he had built 51 years earlier.

hills. During the two years that he continued to mine, he lived at the millsite in a cabin that they had built. His partner's interest in mining began to diminish, and Oran found himself doing most of the work. Soon the partnership was dissolved and McGinnis left the area.

Booth turned the millsite over to Bill Keys who filed on it, and he soon began helping Mr. Keys with a variety of projects at the ranch. He told me, "Keys had a lot of equipment then, but he wasn't a good mechanic. I never knew anybody who worked with engines as much and had so much trouble with them. I had done a considerable amount of work with engines, so it was no problem for me.

"So I'd do a half days work on something, or I'd do a couple of days. There was never any money, but when I needed equipment of some kind, I always came to see Keys and I never had the man tell me no."

When Bill Keys decided that the older children, Willis and Virginia needed more education than could be provided by their mother, he asked Oran if he would teach them. Because his prior experience had been in junior high school, Mr. Booth drove to San Bernardino to see the Superintendent of Schools and to find out the requirments for teaching grammar school, as well as to get the necessary books and supplies.

In a 1978 conversation with me Oran remembered, "The school building was a tent. It had a wooden floor with a wall up so high with a tent on top. Later the tent went to pot, and Keys covered it with old corrugated iron. It was eight by ten or something. We had little tables there for the kids to write on and put their books on. There wasn't a big class, so we didn't need much room. I always started the school at nine o'clock, and we only lasted until noon. We studied a little bit of reading, a little bit of writing, and what one, two, and three is.

"These kids didn't know how to play. We'd have a recess, and I'd show them how to play baseball, to play hide and seek, and some of those games, because they'd played with no other kids. They'd had no form of training at all. So I just tried to start them out."

After six months in the classroom, Oran resumed his mining pursuits full time and prospected in the Gold Park area. He also worked for Bill by doing custom milling at the Wall Street mill where Bill had moved the two-stamp mill from Pinyon Wells. In 1933 he filed on a 80-acre homestead south of Twentynine Palms, built a cabin, and dug a well.

During the war he turned from mining to construction working at Camp Hahn and Camp Irwin. Afterwards he worked at a succession of jobs as a machinist, welder, and mechanic in Long Beach and Wilmington returning from time to time to his cabin in Twentynine Palms.

After his retirement in 1962 he returned here permanently continuing his long and close friendship with Bill Keys and his family, as well as working at his Paymaster Mine south of town, a gift from Bill.

Only in his last few years when his eyesight began to fail, did Oran give up driving and his gold-mining pursuits. Despite this impairment, he still enjoyed working with a variety of equipment and machinery on his homestead property. Mr. Booth was a charter member and long-time supporter of the Twentynine Palms Historical Society, and a 72-year member of the Masons.

He was in good health until a fall and broken hip sent him to the hospital where he died on December 6th, 1994 at the age of 93. His funeral was held the same day as the dedication of Joshua Tree National Park, an area he had seen designated as a national monument in 1936.

On a personal note I miss visiting with him and hearing him recount his early days in the desert and to answer questions on techniques of mining. Only by talking with him at length would one realize that beneath his quiet manner was a well-read individual who had a tremendous knowledge of history and science that he was only too willing to share in conversations with others. He possessed the special qualities and experience that make a great teacher. The Keys family certainly appreciated these virtues in their life-long friend.

Oran Booth at the book signing for the publication of Art Kidwell's *In the Shadow of the Palms, Vol. 1* on December 20, 1986, at the 29 Palms Branch Library.
Photo by and courtesy of Ada Hatch.

DELLA and HOWARD DUDLEY

Of the four teachers who lived and taught at the Desert Queen Ranch, Della Dudley was there the longest. She also made a significant contribution to the education of Pat and Phyllis Keys as well as to the children of neighboring homesteaders and miners who attended the Desert Queen School during its last seven years. With a student body that varied from as few students as six to a dozen or so, Mrs. Dudley brought a wealth of knowledge and experience to the isolated desert location.

Della Augusta Williams Dudley and her husband, Rev. Howard Edward Dudley came to the Desert Queen Ranch in 1937. They had met when they were students at Dennison University, Ohio. Howard graduated in 1899 then went on to study for the ministry.

In the meantime Della became a school teacher, and after Howard's graduation from Rochester Theological Seminary in 1902, he returned to Ohio, was ordained as a minister in Springfield, and persuaded Della to marry him. Two days later they left for Burma where their four children were born, and where they spent thirty-three years as missionaries with the American Baptist Foreign Mission Society. Here Howard supervised the churches and mission schools in Meiktila, while Della served as superintendent of a mission school there and was the author of two textbooks for schools in Burma.

After returning to the States in 1935 and not ready for retirement, Della accepted an unusual teaching assignment from the San Bernardino County School System who sent her to Bill Keys' ranch in the fall of 1937. She and Howard would live at the ranch until the summer of 1942 when the younger Keys children were to move temporarily to Alhambra with their mother to attend junior high and high school there.

During their tenure at the ranch, Della was the teacher, while Howard helped her in the classroom and led the students in their morning prayer. He also helped Bill Keys with chores around the ranch.

Della Dudley's dedication to teaching as well as her innovative methods well prepared her students for life in the outside world far from the isolation of their desert school. Through her guidance, her students not only studied the same curriculum that they would had they been in a larger city school, but they also learned about the uniqueness of the desert environment in which they lived. Class essays and artwork reflecting desert themes were published in the little school's own newspaper, *THE DESERT QUEEN HERALD*.

Field trips away from the school were also part of Della Dudley's teaching curriculum. A May 1941 article in the *LOS ANGELES EXAMINER* detailed one trip by six students from the school who visited the newspaper "not as sightseers but also as fellow members of the press."

The article said that the Desert Queen School students "write, edit and print their own school newspaper on a press they built themselves at a cost of $1 and considerable effort. It's not all the 3-Rs at Desert Queen, as is proven by the fact that the youngsters themselves painted their schoolhouse, built necessary bookcases and lockers and still found time to turn out a book of poetry of the desert and to study art and sketching under Victor Low, noted New York artist, who admittedly fell in love with those youngsters on a recent visit to the Park."

When Mrs. Dudley was notified that a story written by Marion Headington, one of her students, was accepted by radio station KFI, she gave her other students the next day off and left immediately with Marion late that afternoon to spend the night in Riverside in order to be at the radio station for the 7:30 a.m. broadcast the next day.

In 1940 the students wrote a composition about their school which was published in the *JUNIOR RED CROSS NEWS* who "selected it because of its vivid and youthful description of desert life in and about Twentynine Palms." When Della won a cash award for an essay she had written on "How I Use My Environment in Teaching," she donated the entire cash award from the teacher's magazine, *THE INSTRUCTOR,* in the name of Desert Queen School to China Relief. Her students also sent toys that they had made out of wood for young people in China.

Ensuring that her students got a good education was not the only concern faced by Della Dudley. The San Bernardino County Superintendent of Schools controlled the budget for the ranch school, and Mrs. Dudley frequently wrote to C. Burton Thrall requesting necessary operating funds sometimes out of the ordinary. The following excerpts from some of her letters reveal another side of this no-nonsense lady - one who could exhibit a great sense of humor in dealing with a county bureaucracy ninety miles away:

Della and Howard Dudley with Desert Queen Ranch students outside their school building about 1940.

Sept. 14, 1938
Dear Mr. Thrall,

I spoke to you concerning two more children who have entered the Desert Queen's realm and seek transportage currency! The first of my subjects is Sherman Randolph, first grade. His father works in a hardware store in Los Angeles but has asthma badly. They have homesteaded out here so the wife stays to look after the ranch. She is a mighty plucky little woman the way she drives over this desert in an old rattletrap of a car. I know that they are pretty hard pressed and that she really needs the 25 cents a day for gas. I doubt if that covers the amount spent. I do hope you can help her out.

The next one is Richard Auclair, fifth grade. His father is working over at the Desert Queen mine with small chances of getting much out of it. Mrs. Randolph drives about twenty miles a day to bring and take the small boy and Mrs. Auclair about 32 miles.

*　　*　　*

Sept. 26, 1938
Dear Mr. Thrall,

I wanted to speak to you about the well out here. It is in a very unsanitary condition due to that flood last winter. The well was filled with debris and until school closed we were using shallow wells in the bed of the stream. Now they are dry and we have to go back to the well. At the bottom is about two feet of muck. Then the engine with which we used to pump water is also submerged in that mess so we can use the old well-sweep. It was a blow from that well-sweep that eventually killed Ellsworth Keys, and I might also say that our youngest boy, Noel, was hit on the head by it and got married within a week. Bathing except in a tea-cup of water is almost a necessity. I now use perfume and powder to cover up.

As you know the Keys are pretty short on cents. You might say they ain't got none at all. I am pretty scarce myself no matter how you spell it. If you can send me $50 or $60 or.... (Don't be afraid just hold my hand tight - I can't go much higher!), so I'll start getting a man to work out here. Mr. Keys says he can't do it alone and also there will be the fixing of the machine. The school has to drink this water, and I really suffer from the lack of it.

*　　*　　*

Oct. 15,1938
Dear Mr. Thrall,

We got two dead rats out of the well last week. I think they died of varicose veins. I'll be glad to have the thing cemented so it can't be a suicidal brink for despondent critters.

*　　*　　*

Oct. 29,1938

Used furnishings supplied for school 1937-38: 6 chairs, 4 tables, 1 sink, 1 rug, linoleum, 2 bookcases, dishes, cooking utensils, curtains. Complete for $60.00

*   *   *

Almost two months later, Della Dudley received an answer to her letters:

Nov. 12, 1938
Dear Mrs. Dudley:

Your letters have been received and appreciated, but for several weeks I have been decidedly "under the weather", and my correspondence has accordingly been neglected.

Unfortunately the demands upon our Unapportioned Elementary School Fund have been much greater than anticipated, and we have not had the available money to pay you for the school furnishings or to assist Mrs. Randolph in lieu of furnishing transportation for her son. Please be assured that we shall do our best, but we do not know just how good or how soon this will be.

Very truly yours,

C. Burton Thrall, County Superintendent of Schools

After the school closed in 1942, the Dudleys moved to Ontario and later to Alhambra, where Howard died in December 1964. Della died at the home of her daughter, Emma Truesdail in North Carolina the following December.

One of their four children, Howard, became a teacher after he retired from a military career. Several of her grandchildren are also teachers. Emelyn, daughter of Howard and Della's eldest son, Richard, continues the family teaching tradition by providing quality bilingual education to her elementary students in Watsonville, California, and is writing a book on her remarkable grandparents.

The Desert Queen School today.